W9-DFV-797

A GUIDE
TO ALASKA'S
EDIBLE HARVEST

JANICE J. SCHOFIELD

Alaska Northwest Books™
Anchorage • Seattle • Portland

Dedicated to Spirit, weaver of oneness,
with gratitude for the wild plants that rejuvenate our cells and spirits
and reawaken our connections to earth and each other.

A portion of the proceeds from the sale of this book benefits Kachemak Heritage Land Trust, a nonprofit corporation dedicated to preservation of critical habitat in the Kachemak region. For additional information on KHLT, write P.O. Box 2400, Homer, AK 99603.

DISCLAIMER: Though the plants described in this book have been traditionally used as food or medicine, positive species identification in the field is the reader's responsibility. If identity is questionable, do not gather or ingest a plant. Neither the author nor the publisher is responsible for allergic or adverse reactions individuals may experience from wild foods, nor do they claim that the techniques described in this guide will cure any illness.

Third printing 1997

Library of Congress Cataloging-in-Publication Data
Schofield, Janice J., 1951-
    Alaska's wild plants : a guide to Alaska's edible harvest / by Janice J. Schofield.
      p.   cm.
    Includes index.
    ISBN 0-88240-433-4
    1. Wild plants, Edible—Alaska—Identification.   2. Wild plants, Edible—Alaska—Therapeutic use.   3. Cookery (Wild foods).
I. Title.
QK98.5.U6S36   1993
582'.06'309798—dc20   92-35851
                                 CIP

Managing Editor: Ellen H. Wheat
Editor: Carolyn Smith
Copy Editor: Alice Copp Smith
Designer: Cameron Mason

All photographs are by Janice J. Schofield unless otherwise noted.
Cover photos: FRONT: *Fireweed on the Alaska landscape*, John W. Warden. BACK: *Wild rose*, Janice J. Schofield.

Alaska Northwest Books™
An imprint of Graphic Arts Center Publishing Company
2208 N.W. Market Street, Suite 300
Seattle, WA 98107
800-452-3032

Printed on acid- and elemental-chlorine-free recycled paper in the USA

CONTENTS

GRASSY MEADOWS AND FOREST OPENINGS

MARSHES, PONDS, AND WETLANDS

POISONOUS PLANTS

COOKING WITH WILD PLANTS 81

ALASKA'S EDIBLE HARVEST

Plants, like people, live in communities and share commonalities. Plants that flourish together share an affinity for certain soils, lighting conditions, moisture, salinity, or altitude. For this reason, this book is organized by habitat. Learn to recognize habitat and you simplify the plant identification process. Be aware that some plants are highly adaptable and can be found sprinkled in many habitats, but concentrations occur where growing conditions are optimal.

Sea and Sandy Shores include gravel beaches, tidal marshes, and rocky intertidal and subtidal zones. Here many coastal plants including beach greens, goosetongue, and roseroot have developed fleshy stems and leaves in order to withstand desiccation in this salt-kissed environment. Algae, such as bladderwrack, have adapted to the extremes of immersion in sea and exposure to glaring sun.

Gardens, Lawns, and Disturbed Soils are our most familiar habitats. The hard-packed soils of roadsides and driveways and the fertile soils of kempt fields are home to our most enthusiastic plant volunteers. Examples include cosmopolitan wanderers like plantain, dandelion, and horsetail—herbs more nutritious than our garden vegetables but commonly scorned as "weeds."

Forests and Open Woods range from the dense temperate rain forests of coastal Southeast and Southcentral Alaska to the spruce-birch-aspen forests of the Interior. Spruce-hemlock forest soils are often acidic and hospitable to devil's club, ferns, and berries.

Moist Places along stream banks, rocky creeks, and bluffs are prime environments for water-loving wildflowers, like spring beauty.

Tundra and Dry Places include the vast treeless regions of the Arctic, lushly carpeted with lingonberries, Labrador tea, and cloudberries, as well as the dry slopes where serviceberries flourish.

Grassy Meadows and Forest Openings are open spaces often ablaze with columbine, fireweed, and showy wildflowers.

Marshes, Ponds, and Wetlands are environments in which mare's tail and cattail thrive in standing water, and bog cranberry weaves through spongy carpets of sphagnum moss.

Within each habitat section, plants are arranged in alphabetical order by common name. Though useful in everyday conversation, the common names of plants can vary greatly, even within Alaska. Botanical nomenclature is essential for obtaining additional information on specific plants. The genus and species names used in this book follow Eric Hultén's *Flora of Alaska and Neighboring Territories* (Stanford University Press, 1968). Family names are updated to reflect current usage, naming the family after its largest genus. For additional identification details, drawings, plant data, photographs, and an extensive bibliography, consult my comprehensive guide, *Discovering Wild Plants: Alaska, Western Canada, the Northwest* (Alaska Northwest Books, 1989).

I have included a few recipes using wild plants. They are intended as guidelines for spurring your own creativity. Check the glossary at the back of the book for unfamiliar terms. For herbal classes, remedies, seeds, and products, look to the Herbal Directory.

A WORD OF CAUTION. Be certain to review the "Caution" sections carefully. Some plants, such as cow parsnip, can cause dermatitis; others, like red elder, have both edible and toxic portions. When eating any new food for the first time, consume a small amount only. Be sensitive to the effect on your body, and discontinue use immediately and seek medical attention if you experience adverse reactions or allergies. Above all, be positive of identification: A nibble of poison hemlock can have dire consequences.

HARVESTING WITH AWARENESS. Many plant-gathering traditions relate to entering the field or forest with an attitude of thankfulness. Individual expressions of gratitude for nature's bounty can range from planting seeds or sprinkling cornmeal where roots are dug to leaving one's saliva on a rock or singing songs. I find that giving to the plants before gathering increases my focus and awareness.

Growing native plants in your own yard allows you to observe them intimately in all phases of growth. Grow from seed or collect on land slated for development. Be sure to note the soil and light conditions and try to duplicate them in your garden.

Off-limits to foragers are Alaskan state, national, and municipal parks. Harvesting is allowed on state land not designated as parkland, provided that you collect 50 feet back from the highway. In national forests, stay 200 feet back from established trails, roads, and campgrounds. Ask permission to harvest on private land.

Overharvesting, as well as nonharvesting, has been responsible for the loss of many species. Dena'ina tradition, as recorded in Priscilla Russell Kari's *Tanaina Plantlore* (National Park Service, 1987) warns that "if plants that can be eaten aren't gathered and used, there will be less the following year. If this continues, they will all disappear." Responsible foraging enhances the health and productivity of the plant communities and slows natural succession, such as that from field to forest. Begin, when digging roots, by collecting one out of ten roots from productive patches. Monitor your impact by returning year after year to a particular gathering area, and expand your harvesting quotas as appropriate for each species. Some plants may defy depletion, but carefully observe the vigor of the less common species. In the growing tide of herbal interest, many herbs have been "loved" to extinction.

Gather plants in clean areas, away from busy roadsides and toxic sprays. Take only what you can use. For year-round use, dry herbs in a warm, shady, well-ventilated space. (An exception are sea vegetables, which often mold unless quickly sun-dried.) Store herbs in a dark place in airtight containers. Label and date herbs. Storage life is generally six months to a year for green, leafy herbs and one to three years for roots. Supplement these guidelines by comparing the herb's color, taste, odor, and effectiveness to those qualities when it was first dried.

BEACH GREENS
Honckenya peploides   Pink family (Carophyllaceae)

Mats of bright green "beach greens" are a familiar sight on many Alaskan beaches. The trailing stems grow to 2 feet long, with smooth, fleshy leaves in opposite arrangement. The small greenish-white flowers are followed by globular seed capsules.

Derivation of name: *Honckenya* honors 18th-century German botanist Gerhard Honckeny; *peploides,* from the Greek word for a cloak, describes the way the leaves wrap the stems and nearly hide the flowers.

Other names: sea chickweed, seabeach sandwort, sea purslane.

Range: sandy shores from Southeast Alaska to the Arctic.

Harvesting directions: Young leaves and shoots are prime before flowers appear; wash well to remove the grit and sand that may be trapped by the cloaklike leaves.

Food use: Beach greens bring mixed reviews from foragers. Some compare the flavor to oysters. Others, perhaps those who don't like oysters, prefer milder greens. I like adding the young leaves to sea-side vegetable dishes, including stir-fries, soups, and salads (especially with vinegar and oil dressing). As the flowers develop, the leaf flavor intensifies, but the plant can be quite useful on late-summer camping trips, especially when blended with goosetongue, beach peas, and lovage. Fermented beach greens are a favorite on the Seward Peninsula and are said to be similar to sauerkraut.

Medicinal use: Beach greens were consumed by circumpolar explorers seeking vitamin C–rich greens to cure scurvy. The leaves are also a good source of vitamin A.

Other: Though beach greens can form a comfortable mat for your sleeping bag, you should be aware that they are also a favorite food of black bears.

BEACH PEA

Lathyrus maritimus, aka *L. japonicus*   Pea family (Fabaceae)

photo by Norma Wolf Dudiak

Beach peas are listed as toxic in some plant guides, due to the presence of a cumulative toxin (which also occurs in garden peas). Unlimited pea consumption can lead to lathyrism, characterized by partial or total paralysis. This irreversible condition has occurred during famines, when peas were the sole food source. Another challenge is identification, as some peas *are* toxic, even in small quantity, and the family is a large one! Beach pea is specific to gravel beaches, bears a typical reddish-pink to purple "pea" blossom with wings and keel, has smooth unjointed pods, and has tendrils at the end of the stem. Though caution is well deserved in this often-confusing family, beach peas, properly identified and eaten in moderation—as one should eat all foods—are safe and nourishing.

Derivation of name: *Lathyrus* is from the Greek *la thoursos,* meaning "something exciting." *Maritimus* indicates the plant's habitat.
Other names: purple beach pea, seaside pea, raven's canoe.
Range: Southeast Alaska shores to Icy Cape in the Chukchi Sea.
Harvesting directions: Pick young shoots when under 10 inches high. Pick pea pods when bright green.
Food use: Young beach-pea greens can be steamed as a potherb or stir-fried. Young tender peas can be eaten whole like snap peas. Mature peas, removed from the pod, are a tasty raw snack or cooked vegetable. (See Caution, following.) Beach peas are high in protein and vitamins A and B.
Medicinal use: The Chinese prescribe cooked beach-pea greens for toning and nourishing the intestinal tract and urinary organs.
Other: Though American Indians commonly fed marsh vetchling, *Lathyrus palustris,* to their horses, edibility for humans is questionable. Marsh vetchling thrives in swampy ground and has butterfly-shaped leaves (stipules) at the base of its branched stems.
Caution: Positive identity is crucial. If in a survival situation, do not subsist solely on beach peas; nervous disorders and paralysis can occur from overconsumption.

BLADDERWRACK

Fucus spp.   Brown algae (Phaeophyta)

Bladderwrack is an easy-to-recognize sea vegetable, found along the ocean shore attached to rocks. The olive-brown blades, which grow to 18 inches in length, divide in twos and have prominent midribs. The distinctive inflated tips contain mucus; these bladders also contain the male and female gametes (eggs and sperm) that have propagated these algae for over 400 million years.

Derivation of name: *Fucus* is from the classical Greek *phykos*, meaning "seaweed."

Other names: wrack, paddy tang, rockweed, old man's firecrackers, kelp popping weed, *tuhbet*.

Range: Southeast Alaska to the Chukchi Sea.

Harvesting directions: In spring to early summer, collect the inflated tips and tender ends of fronds. Clip above the small disklike holdfast that anchors the sea vegetable to rocks.

Food use: Nibble bladderwrack raw on beach hikes. It is a rich source of trace minerals and protein. Add tender tips and fronds to stir-fries, soups, sauces, and egg dishes. Steam with seafood to heighten flavors. Pour boiling water over sun-dried *Fucus* for a tasty drink or soup starter. Dry and grind to use as seasoning.

Medicinal use: Medical studies at McGill University confirm that sodium alginate in *Fucus* binds with strontium 90 (which is present in nuclear fallout) and removes the radioactive substance from the body. *Fucus* tea is a cleansing tonic. Herbalists also recommend bladderwrack in teas, tinctures, and capsules for stimulating the metabolism, regulating thyroid function, and relieving goiter. The mucilaginous tips, similar in texture to aloe vera, are used to treat burns, corns, and skin tumors. Add bladderwrack to footbaths for sore feet and swollen ankles, and to liniments for rheumatic pain.

Other: The powdered herb is a dietary supplement for both pets and humans. Bladderwrack is a potash-rich fertilizer for potatoes and cole crops. Burn dry bladderwrack with alder to smoke salmon.

BULL KELP

Nereocystis luetkeana   Brown algae (Phaeophyta)

photo by Norma Wolf Dudiak

Tangles of bull kelp are a familiar hazard to boaters and a favorite resting place for sea otters. These giant algae can grow to 200 feet in length in a single year! A branchlike holdfast attaches bull kelp to the sea floor. At the surface, an inflated bulb floats long, narrow blades.

Derivation of name: The genus name is from the Greek *nereo,* "sea nymph," and *cystis,* "bladder"; *luetkeana* honors a Russian sea captain.

Other names: bull whip kelp, bulb kelp, giant kelp, sea kelp, horsetail kelp, sea otter's cabbage, *tutl'ila.*

Range: Southeast Alaska to the Aleutian Islands.

Harvesting directions: Bull kelp is prime from April to June. Collect plants attached to the ocean floor. Peel the long hollow stipes (stems) immediately after harvesting. Non-boaters can check the beach immediately after storms for freshly uprooted kelp. Use older plants for garden mulch.

Food use: Dry bull kelp blades and grind as a table seasoning. Cut the peeled stipe into "kelp rings" and pickle or marinate for superb snacks. Stuff the kelp bulb with meat or vegetarian stuffing and bake. Wrap fresh kelp blades around fish, cover with foil, and bake until tender; serve the kelp as a vegetable. Kelp is low in fat and a good source of trace minerals, protein, and carbohydrates.

Medicinal use: Herbalists use kelp capsules, teas, and seasoning for overactive thyroids and to speed the healing of fractures.

Other: Kelp is an internal and external beauty aid for the skin and nails. To make candles, fill kelp bulbs with a candlewick and hot wax; when the wax is set, discard the kelp. Kelp stipes and bulbs are used as horns and rattles by musical campers.

Caution: Some individuals experience gas from ingesting kelp. Goiterlike symptoms may result from excessive use.

DULSE
Palmaria mollis (formerly *Rhodymenia palmata*)
Red algae (Rhodophyta)

If you're eating sea vegetables for the first time, try dulse. Dulse "potato chips" (directions below) are popular with kids and adults because of their salty taste and crunchy texture. The plants also make a good trail snack eaten raw. Look for purple-red blades, 4 to 12 inches in length. Raw dulse is somewhat rubbery.

Derivation of Name: *Palmaria* is from the Latin for "palm of the hand"; *mollis* means "soft."

Other names: dillisk, red kale, Neptune's girdle, water leaf.

Range: Southeast Alaska to the Aleutian Islands.

Harvesting directions: Dulse is prime from April to June. Clip fronds above the holdfast and pick clean of shells. If you choose to rinse the dulse, use salt water to minimize deterioration.

Food use: Dulse can be nibbled fresh from the beach, or added to stir-fries, soups, and sandwiches. It is high in protein, calcium, phosphorus, potassium, magnesium, trace minerals, and vitamins A and B. Keep a shaker of powdered dry dulse handy to add flavor and nutrition to your foods. Add chopped dry dulse to spreads and dips. A favorite snack is a cracker spread with cream cheese and generously topped with dulse. For dulse "potato chips," quickly stir-fry dry dulse in a few drops of hot olive or sesame oil, stirring constantly until crisp. (Be careful not to burn the dulse.)

Medicinal use: Herbalists recommend mineral-laden dulse tea for menstruating and lactating women. The sea vegetable is said to help regulate thyroid function. Dulse is added to salves for herpes simplex.

Other: Eat dulse to promote growth of your hair and nails. I enjoy a rejuvenating dulse bath after airline travel; during long journeys, I consider sea vegetables an essential snack.

GLASSWORT

Salicornia spp.   Goosefoot family (Chenopodiaceae)

From the coasts of Alaska and Florida to northern Europe and southern New Zealand, you'll find species of *Salicornia*. Alaskan specimens are generally 2 to 6 inches in height. The fleshy stems are jointed and may vary in hue from green to red. Flowers are minute, in groups of three, hidden in the joints.

Derivation of name: *Salicornia* translates as "salty horn," describing both the taste and the shape of the herb.

Other names: beach asparagus, pickle plant, saltwort, samphire, chicken claws, pigeonfoot.

Range: Southeast Alaska and Kenai Peninsula to Anchorage.

Harvesting directions: Collect the tender aboveground portions of glasswort. Pick only a few stems from each plant so that the patch will continue in abundance.

Food use: Glasswort is tasty raw or cooked. Add glasswort tips to salads or steam as a potherb. Its salty flavor makes it ideal for seasoning campfire soups and cooked dishes. The whole plant can be steamed as an asparagus substitute and eaten like an artichoke leaf. Dip in butter or curry mayonnaise and then drag the stems between your teeth to remove the tender portions.

Medicinal use: According to Nicholas Culpeper's 1826 *Herbal*, glasswort juice or powder is cleansing and useful as a diuretic for water retention (though harmful in excess). Culpeper also advocated the powder for drying running sores and curing ringworm.

Other: Ashes from burned glasswort were combined with pulverized stones and melted to make glass. Ashes have been used commercially in soap-making. According to 16th-century English herbalist-physician John Gerard, burning the plants drives away serpents. It would be more useful in Alaska if it kept bears out of camp.

GOOSETONGUE
Plantago maritima   Plantain family (Plantaginaceae)

Goosetongue is a coastal plantain species, renowned for its fine flavor, but it demands attentiveness on the part of foragers because of its similarity to toxic arrowgrass (see Caution, following). Goosetongue has compact flower stalks (generally 6 to 12 inches in length); its yellowish stamens stick out and dance in the breeze. Leaves are fleshy with a pleasant, somewhat salty, flavor.

Derivation of name: *Plantago* means "sole of the foot." *Maritima* means "maritime."

Other names: seaside plantain, narrowleaf plantain, ribwort, sheep's herb.

Range: seacoasts and salt marshes from Southeast Alaska to the Aleutians and on the southern Seward Peninsula.

Harvesting directions: Goosetongue is prime from spring to early summer. Later in the season, look for young tender leaves growing in the center.

Food use: Goosetongue is one of the most popular seaside greens, both raw and cooked. Chop the leaves and add to salads, casseroles, spanakopita, and stir-fries. Steam greens as a potherb; try them topped with garlic butter. Can or freeze for winter use.

Medicinal use: Apply mashed goosetongue to relieve itching mosquito bites.

Other: Place goosetongue seed stalks in bird cages.

Caution: Occasionally some individuals are reported to experience allergic reactions to *Plantago* species. Foragers should use care in identification to differentiate from arrowgrass, *Triglochin maritima*, which often grows in proximity to goosetongue. Mature arrowgrass bears stalks 1 to 2½ feet high and has greenish-white flowers; the acrid-tasting leaves contain cyanide. See Poisonous Plants.

LOVAGE

Ligusticum scoticum   Parsley family (Apiaceae)

Observe the young leaves of lovage and you'll often notice a characteristic red line around the margin. Stems divide in three, each bearing three leaflets. Crush a leaf and notice its distinctive scent. Flowers are white to light pink in hue, arranged in umbels with 7 to 11 rays. Stems are reddish-purple at the base and grow to 2 feet high.

Derivation of name: *Ligusticum* means "from Liguria," an Italian province; *scoticum* refers to Scotland.

Other names: Scotch lovage, beach lovage, sea lovage, wild celery, *petrushki, bidrushga, pidrushga, tukaayuk.*

Range: sandy beaches from Southeast Alaska north to Nome.

Harvesting directions: (See Caution, following.) Leaves are mildest and most tender before flowering; later in the season, look for tender new growth in the center. Collect seeds in late summer.

Food use: Chop lovage and sprinkle on crackers with cream cheese; garnish with edible flowers or wild berries for an attractive hors d'oeuvre. Lovage is high in vitamins A and C. Add fresh or dry leaves to fish, chowders, spaghetti sauce, and holiday stuffings. Seeds season stews, breads, and sausage.

Medicinal use: Traditionally, lovage seeds are a carminative tea (for expelling gas), a digestive aid, and a flavoring agent for medicines.

Other: To remove odors after handling garlic or fish, rub the crushed leaves between your palms. Add a large handful of lovage to a bucket of hot water for a fragrant and deodorizing sauna bath; lovage baths are traditional for drawing romance to your life!

Caution: The parsley family is large and diverse, containing prime edibles like lovage, potent and strong-tasting medicinal herbs like angelica, and deadly plants like poison hemlock (*Cicuta* spp.). All bear umbrellalike flower clusters (umbels) and require positive identification of species. See Poisonous Plants.

NORI
Porphyra spp.   Red algae (Rhodophyta)

Floating in water, nori resembles a long transparent rubber glove. Look closely and you'll see reddish-purple blades up to a foot in length, attached to rocks by a small holdfast. On warm days, at low tide, nori looks like a black smear on exposed rocks, but don't let its unappetizing appearance deceive you. This favorite of sea vegetables commands a premium price, for good reason.

Derivation of name: *Porphyra* is from the Greek for "purple."
Other names: laver, purple laver, red laver, *thalkush.*
Range: Southeast Alaska to the Aleutians.
Harvesting directions: Collect nori at low tide, gathering the entire plant except the holdfast. Rinse with salt water to remove sand and grit.
Food use: Munch nori raw, or chop fine and add to salads, stir-fries, and soups. Nori is a rich source of protein, iron and other trace minerals, and vitamins A, B, and C. Dry nori can be nibbled as a snack, or ground and added to flour, vegi-burgers, or seasoning mixes. Steam seafood with nori to enhance flavor. Roll fresh nori into a ball and dust with flour or ground oats; toast on a skillet lightly coated with oil. This is a traditional breakfast food in Wales. For a special snack, dip nori in an egg-milk mixture and then roll in pancake mix and fry in oil; the flavor is similar to that of fried oysters.
Medicinal use: Nori has been used in treating goiter and scurvy.
Other: Nori is the alga used in making the popular Japanese treat sushi. The sheets of nori, used to wrap the vinegared rice, are prepared through a process similar to paper-making. The Japanese are said to consume 9 billion nori sheets annually.
Caution: An excess of nori can cause gas and stomach upset.

ORACH

Atriplex spp.   Goosefoot family (Chenopodiaceae)

Orach, like lamb's-quarter, is a member of the goosefoot family. Orach seed develops within pairs of clasping leaves. Stem leaves are variable with species, ranging from triangular to lanceolate-oblong. Lower leaves are opposite, upper leaves alternate.

Derivation of name: *Atriplex* may come from the Latin roots for "dark" and "fold," describing the plant's clasping leaves.

Other names: saltbush, seascale, shadscale, sea purslane.

Range: coastally from Southeast Alaska to the Alaska Peninsula and on the southern and northern Seward Peninsula.

Harvesting directions: Collect the tender aboveground shoots and leaves in spring. Gather seeds from late summer to early fall.

Food use: Orach is edible raw or cooked. Add to campfire soups, egg scrambles, and salads. Roast seeds, or grind to flavor and extend flour supplies. Boiled greens are a potherb; Southwest American Indians use the cooking water in corn dishes. Ashes from the burned plants have been used as a baking-powder substitute.

Medicinal use: Mash leaves and roots and apply as a poultice for insect bites. Traditionally, leaves are boiled with vinegar and applied externally for gout.

Other: Orach is cultivated in Africa and Eurasia. *Atriplex hortensis*, dubbed New Zealand spinach, is becoming increasingly popular with American gardeners. *Atriplex* is a hardy herb with more than 100 species worldwide, which adapt to the extremes of salty Arctic beaches and alkaline deserts.

Caution: Orach can absorb selenium from the soil. Though poisoning is unlikely in Alaska, livestock-poisoning has occurred in selenium-rich areas such as South Dakota and Wyoming.

OYSTERLEAF
Mertensia maritima   Borage family (Boraginaceae)

The distinctive blue-gray cast of oysterleaf leaves makes this plant easy to recognize. The bell-like blossoms vary in hue from pink and blue to white; flowers are similar to those of garden comfrey (another member of the borage family). Stems trail on the ground and rise up at the tip. Another identifier is the flavor; take a taste and you'll discover the origin of its common name.

Derivation of name: *Mertensia* honors botanist Franz Karl Mertens; *maritima* refers to the habitat where this plant thrives.
Other name: oyster plant.
Range: beaches from the northern Panhandle to the Arctic.
Harvesting directions: In spring, gather the basal leaves. Later in the season, harvest stem leaves and flowers. Though oysterleaf is most tender in May and June, leaves are palatable throughout the growing season.
Food use: Nibble tender oysterleaf as a beach snack. Use as a base for salad or sandwich filling. Add to quiche and egg dishes, as well as chowders and soups. Flowers are an edible garnish.
Medicinal use: In northern Europe, oysterleaf blended with fennel and honey water is a traditional remedy to soothe coughs.
Other: Oysterleaf can vanish due to overgrazing by wandering stock. An Alaskan friend has fenced her seaside property to preserve oysterleaf, goosetongue, and other beach edibles for her dining pleasure.

RIBBON KELP

Alaria spp.   Brown algae (Phaeophyta)

photo by Norma Wolf Dudiak

Ribbon kelp is easy to recognize with its smooth olive-green to brown blade, 3 to 9 feet long, and its distinctive flattened midrib. At the base, between the holdfast and the main blade, are two opposite rows of smaller winglike blades. *Alaria marginata* is often marketed in natural-food stores as *wakame* or *Alaria*, but is readily available, for free, on many Alaskan beaches. Though all *Alaria* species are reported safe, some species are less appetizing than others because of their hairy texture or blistered midribs.

Derivation of name: *Alaria* means "winged" and refers to the winglike sporophylls between the holdfast and the large main frond.
Other names: *wakame,* wing kelp, tangle, honeyware, murlins.
Range: Southeast Alaska to the Bering Sea.
Harvesting directions: Ribbon kelp is prime from April to June. Clip ribbon kelp above the winglike lower sporophylls. Discard tattered edges. Pick clean of shells.
Food use: Nibble ribbon kelp fresh. The central midrib has a delightful crunchy texture and mild flavor. Ribbon kelp is an ideal snack food—low in calories, and high in flavor and nutrients, especially protein, calcium, phosphorus, potassium, magnesium, trace minerals, sulfur, sodium, and B vitamins. For year-round use, dry the fresh blades. Add a teaspoon of the powdered dry alga to breads and baked goods to increase nutritional value.
Note: The sporophylls (lower winglike portions) are considered a delicacy raw or cooked, but they should be harvested selectively because they regenerate the algae.
Medicinal use: Like bladderwrack, ribbon kelp contains sodium alginate, which removes strontium 90 (a radioactive substance present in nuclear fallout) from the body. Ribbon kelp tea is a low-calorie, high-mineral beverage for dieters.
Other: *Wakame* steams are soothing for the complexion. For a special facial, blend white cosmetic-grade clay with powdered ribbon kelp.

ROSEROOT

Sedum rosea   Stonecrop family (Crassulaceae)

photo by Norma Wolf Dudiak

The name "roseroot" describes the fragrance of *Sedum's* rootstalk. Roseroot is a tidy plant that perches on beach cliffs, thriving in a thimbleful of soil. The pale leaves are spoon-shaped and overlap in a spiral fashion. The early-blooming flowers sit at the top of the stem; look closely and you'll see that individual blooms have four petals and bear pistils only (purple) or stamens only (yellow).

Derivation of name: *Sedum* is from the Latin "to sit"; *rosea* means "rose."

Other names: hen and chickens, rosewort, scurvy grass.

Range: beaches, rocky places, and alpine slopes throughout Alaska.

Harvesting directions: Pick leaves before flowering for the tastiest fare. Campers can extend the harvest season by collecting the young growth on summer stalks.

Food use: Eat roseroot leaves raw, in salads and coleslaws, or steam as a potherb. The greens, which are high in vitamins A and C, blend especially well in egg dishes. Include in stir-fries, soups, and vege-table casseroles. Eskimos ferment the rootstalks and serve them with seal oil or blubber.

Medicinal use: Dena'ina Athapaskans use the leaves and rootstalks in cold teas, sore-throat gargles, and eyewashes. The rhizome is also applied as a poultice for cuts. According to ethnobotanist Margaret Lantis, Nunivak Island Natives serve a decoction of the flowers to relieve stomachache.

Other: Planting *Sedum* in pots and placing them on the rooftop is an old Chinese custom for warding off housefire. I found volunteers for my garden in an eroding beach bluff, hitchhiking to sea in a clump of earth. Each year the plant produces its "chicks," which make great gifts for friends' gardens or rooftops.

Caution: Some sensitive individuals have been known to experience nausea or headache after overindulging in *Sedum* species.

SEA LETTUCE

Ulva spp.   Green algae (Chlorophyta)

The lettuce of the sea is easy to recognize. Look for paper-thin, green, transparent blades growing on rocks at low tide. The broad blades can grow to 2 feet in length, and edges may be ruffled. The common *Ulva fenestrata* develops tiny holes in older blades.

Derivation of name: *Ulva* is a classical Latin name first used by Virgil; *fenestrata* means "containing tiny holes or windows."

Other names: water lettuce, green laver.

Range: Southeast Alaska to the Aleutians and Kamchatka Peninsula.

Harvesting directions: Gather sea lettuce in spring to early summer, when the blades are bright green in color. Pick free of shells. Remove areas bruised by the tide and rocks. If sandy, rinse with cool salt water.

Food use: Young sea lettuce can be chopped for salads, or boiled briefly (one to two minutes) as a potherb. Add to stir-fries and noodle dishes. Sea lettuce is my favorite sea vegetable for a powdered seasoning. Its bright green color and salty flavor make it a colorful and tasty garnish for soups, salads, and even popcorn. To dry sea lettuce, place on a hot sunny windowsill or in a vegetable/herb drier. The thin blades (only two cells thick) dry quickly and powder easily in a blender. Place in a pretty shaker on your table as a reminder to flavor your foods with sea lettuce! As an extra bonus, the seasoning is high in protein, iron, trace minerals, and B vitamins.

Medicinal use: Sea lettuce can be applied as a poultice for treating sunburn.

Other: If you see sea lettuce with white edges, it is because the mature plant has released its free-swimming reproductive cells. Harvest early in the summer before this occurs.

CHICKWEED

Stellaria media   Pink family (Caryophyllaceae)

What gardener isn't familiar with chickweed—the prolific annual that overruns cultivated soil? Take a close look and you'll discover a fine line of hairs running down one side of the stem only. The leaves are arranged in pairs on weak stems that often trail on the ground. Take a nibble, and you'll discover a very mild-flavored green. Look closely at the small white star-shaped flowers; you'll see that the ten petals are actually five divided ones. My husband carefully "sculptures" the lush chickweed patches in the lawn; besides providing green clippings for our salads and soups, they yield seeds for visiting spruce hens and chicks.

Derivation of name: *Stellaria* refers to the starlike flowers; *media* means "intermediate."

Other names: winterweed, satin flower, stitchwort, starwort, adder's mouth, skirt buttons.

Range: most areas of Alaska except north of the Brooks Range.

Harvesting directions: Clip the tender tips of chickweed with scissors throughout the growing season. (Discard the stringy stems.) In mild areas, chickweed can be found blossoming even on winter days. Seeds are also edible, though tedious to collect in quantity.

Food use: Eat chickweed tips raw or steamed lightly. Add to soups, stir-fries, salads, and Wild Herb Pesto (see Cooking with Wild Plants). Chickweed is low in calories and high in copper, iron, phosphorus, calcium, potassium, and vitamins A and C. Juice it with carrots for an energizing drink.

Medicinal use: Add chickweed to herbal salves for burns and cuts and to herbal oils for the skin. Apply chickweed poultices for infections, insect bites, eczema, and psoriasis. This "magic healer" of Euell Gibbons was his favorite in chlorophyll-laden "green drinks."

Other: *Stellaria* is well liked by domestic chickens, rabbits, and pigs. Feed seeds to caged birds. Add chickweed to cosmetic lotions and herbal baths.

CLOVER
Trifolium spp.   Pea family (Fabaceae)

From Alaska to New Zealand, one finds clovers along roadsides and meadows. Though clover is well known as a forage crop for grazing animals, its human use today is less recognized. Clover blossoms have 30 to 40 tiny pealike flowers per cluster; colors vary from white to red. Look for 3 leaflets per stem; consider yourself lucky if you find a 4- or 5-leaved clover.

Derivation of name: *Trifolium* refers to the typical "three leaves."
Other names: common clover, shamrock (*Trifolium* spp.), Alsike clover (*T. hybridum*), white clover (*T. repens*), red clover *(T. pratense).*
Range: most areas of Alaska except north of the Brooks Range.
Harvesting directions: Harvest clover leaves before flowers appear. Pick blossoms at their prime, before they start to turn brown. Roots are best in spring or fall.
Food use: Steam clover leaves as a potherb or add to casseroles, soups, and stir-fries. Though clover leaves may be nibbled raw or added to salads, it is said that cooking them briefly or soaking them in salt water for two to three hours increases digestibility. Dry leaves can be added to baked goods. Dip clover blossoms in batter and fry as fritters. Grind dry blossoms and use as a flour extender. Roots are a traditional food of British Columbia Indians.
Medicinal use: Red clover blossoms are highly regarded by herbalists as blood purifiers. Infusions have been used in traditional treatments for cancerous growths, eruptive skin diseases, hepatitis, and mononucleosis. Add clover to salves and ointments for burns and ulcers. Poultices can ease athlete's foot.
Other: Steep clover blossoms in your bath. They add fragrance, are soothing to the skin, and are a traditional "charm" for attracting abundance. Lore dictates that you gather white clover under a full moon and present it to your lover as a pledge of your fidelity.
Caution: Clover can cause digestive upset if eaten in excess.

DANDELION

Taraxacum spp.   Aster family (Asteraceae)

The cheery yellow composite blossoms and jagged *dent-de-lion* (lion-tooth) leaves are readily recognized worldwide. Stems, when broken, yield a milky sap. Dandelion taproots can penetrate the earth to a depth of 20 feet.

Derivation of name: *Taraxacum* means "remedy for disorder."
Other names: lion's tooth, yellow gowan, priest's crown, blowball, wild endive, swine snout, cankerwort.
Range: throughout Alaska.
Harvesting directions: Pick dandelion salad greens in very early spring, as soon as they appear. As buds begin to form, collect the dandelion crown, between root and bud. Harvest flowers when fully open. Dig roots in early spring or after fall frost and scrub well.
Food use: Though least bitter before flowering, greens are palatable even in summer if marinated. Try a green drink of dandelion, lamb's-quarter, chickweed, and sorrel blended with tomato juice. Fry batter-dipped blossoms as tempura. Flowers are a favorite for making wine, cordials, and dandelion stout. Tender roots can be chopped for stir-fries, or roasted and ground as a coffee substitute. Dandelions are a superb source of calcium, iron, calcium, and vitamins A, B, and C.
Medicinal use: Roots are renowned as skin and liver tonics. Herbalists recommend root decoctions and tinctures to lower cholesterol and high blood pressure, and as a diuretic for water retention.
Other: Add dandelion flowers or floral essence to your bath to relieve muscular tension and to brighten your day.
Caution: Avoid harvesting where weed killers are used.

HORSETAIL

Equisetum arvense   Horsetail family (Equisetaceae)

Horsetail is one of the oldest plants on earth, a remnant of dinosaur days. It has two growth forms, the early brown stalk with a swollen conelike head that releases clouds of spores, and a later green sterile stalk, whose branches lengthen with age.

Derivation of name: *Equisetum* translates as "field horse bristle"; *arvense* means "pertaining to cultivated fields."

Other names: field horsetail, puzzlegrass, scouring rush, pewterwort, jointed grass.

Range: throughout Alaska.

Harvesting directions: Pick the green stalks in spring, while the branches are still pointing upwards. (See Caution, following.)

Food use: Horsetail is used primarily as a mineral-rich tea. Its texture deters use as a potherb; however, Spring Soup (see Cooking with Wild Plants) makes a delicious and highly nutritious meal.

Medicinal use: Horsetail is a traditional poultice for cysts, infections, and bleeding. The mineral-rich infusions are specific for anemia, and for strengthening hair and nails. British herbalist David Hoffman recommends horsetail infusions for prostate troubles, and for bedwetting in children.

Other: Condition your hair with horsetail teas. Spray tea on plants to repel aphids. Scrub pots with mature horsetail.

Caution: Raw horsetails contain thiaminase, a vitamin B–depleting enzyme. Old plants can irritate the kidneys. Internally, use only young plants, and consume as tea or after cooking. The uses above are specific to field horsetail. Be careful not to confuse with wood horsetail *(E. silvaticum),* whose branches are softer to the touch and more feathery in appearance. Many *Equisetum*s frequent wet areas, have conelike spore structures atop the green stalk, and bear few or no branches. Animals that overgraze fresh or dried horsetail can suffer convulsions and loss of muscular control. Vitamin B_1 shots are used as an antidote.

LAMB'S-QUARTER

Chenopodium album   Goosefoot family (Chenopodiaceae)

Lamb's-quarter is a delicious cosmopolitan "weed" that volunteers in many gardens where animal manure is used. Other gardeners who want this nutritious green can purchase seed (see Herbal Directory). Leaves are wedge-shaped and have a whitish water-resistant film. Stems average 1 to 3 feet but can grow to 10 feet. Flowers are inconspicuous, and are followed by abundant seed that self-sows.

Derivation of name: *Chenopodium* translates as "goosefoot." *Album* means "white."

Other names: goosefoot, pigweed, wild spinach, fat hen, fat goose.

Range: Southeast Alaska to Kodiak and the western Yukon and north to the Brooks Range.

Harvesting directions: Collect the entire herb in late spring, the tender leaves in summer. In autumn, collect seeds and leaves.

Food use: Nibble lamb's-quarter as a snack. Add leaves to salads. Greens are high in protein, vitamins A and C, the B vitamins thiamine, riboflavin and niacin, and the minerals iron, calcium, phosphorus, and potassium. Greens freeze well; blanch 1 minute, cool, drain, and package in freezer bags or containers. Dried leaves and seeds, which are closely related to quinoa, are a good addition to winter soups. Seeds substitute for poppy seeds in cakes and breads.

Medicinal use: Lamb's-quarter serves as a poultice for wounds and inflammation. Decoctions are traditionally used as a rub for rheumatism and as a wash for mouth sores.

Other: Napoleon's troops used ground lamb's-quarter seeds to make a black bread.

Caution: Contains oxalic acid, which binds with calcium in the body, forming calcium oxalate crystals that can damage the kidneys.

NETTLE

Urtica gracilis, U. lyallii   Nettle family (Urticaceae)

Nettles are my favorite spring green—bursting with flavor and nutrition. And they're incredibly easy to identify. If you touch one bare-handed, you'll feel a burning sensation! Look closely and you'll notice the soft but stinging hairs on both leaf and stem. Leaves bear coarse teeth and are in pairs that rotate along the stem. Spring plants are often tinged with red. Later in the season, small flowers grow in drooping clusters at the intersection of leaves and stem.

Derivation of name: *Urtica* is from the Latin root *uro*, "to burn."
Other names: Stinging nettle, burning nettle, seven-minute itch, Indian spinach, itchweed.
Range: Southeast Alaska to the Interior.
Harvesting directions: Wear gloves and harvest early spring growth when plants are under a foot in height. Avoid older greens and flowering plants, as they can irritate the kidneys. For year-round use, freeze or dry spring nettles.
Food use: Lightly steam, boil, or stir-fry nettles quickly to tame the "sting." The cooked greens are delectable, like extra-tasty spinach, and are high in protein, chlorophyll, vitamins A, C, and D, and minerals iron, calcium, potassium, and manganese. Be sure to save the cooking water for tea or soup stock. Add powdered dry nettles to bread mixes, seasoning blends, and homemade pasta.
Medicinal use: Iron-rich nettles are an excellent food for anemics, and menstruating and lactacting women. Infusions and tinctures treat eczema. Deliberately stinging oneself with nettles is said to relieve arthritic pain. Nettle liniment is a gentler treatment.
Other: Nettle's strong fibers are woven into nets, string, ropes, and fabric. The herb supplements livestock feed, fertilizes plants, dyes cloth, and yields rennet for cheese-making.
Caution: Wear gloves when handling nettles. Nettle rash can be treated with nettle juice, as well as mashed dock, plantain, or jewelweed.

PINEAPPLE WEED
Matricaria matricarioides   Aster family (Asteraceae)

The ground-hugging wild-chamomile carpet thrives in sunny paths and waste areas. Rub the greenish-yellow rayless flowers between your hands and notice the fresh pineapple scent. Leaves are feathery, finely dissected, on stems 3 to 12 inches high. The plants, which frequent hard-packed soil, may be erect or sprawling.

Derivation of name: *Matricaria* means "dear mother."
Other names: Alaskan chamomile, wild chamomile, dog fennel.
Range: most of Alaska, though rarely in the northernmost Arctic.
Harvesting directions: Pick the flowers in mid- to late summer. Gather foliage throughout the growing season.
Food use: Add flowers to salads and soups. Nibble blossoms as a trail snack. Steep the herb as a tea. I find that the flowers alone make the best tea, but many foragers use both foliage and flowers.
Medicinal use: Alaska Natives use soothing pineapple-weed tea for mothers after childbirth. Infusions can be rubbed on the gums of teething infants. Aleuts and Russians drink the tea for gas pains, stomach upset, and general malaise.
Other: Add pineapple weed to your evening bath to promote relaxation and sleep. To relieve stress or insomnia, indulge in a soothing *Matricaria* bath, sip a pineapple-weed liqueur or tea, and then rest your head on a wild chamomile–filled pillow. Blonds can highlight their hair with a *Matricaria* hair rinse. In the Middle Ages pineapple weed was strewn on floors to freshen the air. Rub flowers on hands to remove strong odors.
Caution: Though it is generally regarded as safe enough for babies, some individuals experience nausea and vomiting from large frequent doses of pineapple weed. Those with sensitive skin also sometimes experience irritation from handling the herb.

PLANTAIN
Plantago major   Plantain family (Plantaginaceae)

photo by Norma Wolf Dudiak

Wherever Caucasian explorers embarked, plantain followed, giving rise to the nicknames "white man's foot" and "Englishman's footstep." Plantain also bears sole-shaped leaves with parallel veins that often lie close to the ground as though stepped on. Its inconspicuous flowers grow on upright spikes; most evident are the yellow stamens that hang from the stem and "hula" in the wind.

Derivation of name: *Plantago* means "sole of the foot."
Other names: cart-track plant, soldier's herb, waybread.
Range: Southeast Alaska to the Brooks Range in disturbed soils.
Harvesting directions: For food use, pick small tender leaves in spring. Collect seeds in late summer by rubbing the mature stalk between the hands and blowing lightly to remove the chaff.
Food use: Add young spring leaves to salads, stir-fries, and steamed vegetable dishes. Add seeds to bread, or use in place of poppy seeds in muffins. Toasted seeds have a flavor similar to artichokes.
Medicinal use: For years I'd heard testimonials regarding plantain's powers in drawing out infection and even shards of glass, but nothing is more convincing than witnessing it firsthand. My hiking companion began our trek with a fishing wound that had been treated at the local hospital with standard antiseptics. Many miles later, an examination of his aching arm revealed blood poisoning streaks running from wrist to elbow. We harvested plantain from the trail, boiled the herb in water, and applied fresh leaf poultices to the wound every two hours. By the next afternoon, all traces of swelling and blood poisoning had disappeared! Since plantain dries very poorly, the herb should be used fresh, or processed into salve or tinctures. Seeds, which are related to psyllium, are a wilderness laxative.
Other: Use plantain in facial steams and creams for acne and in herbal baths for skin rashes. Place leaves in shoes to prevent blisters.
Caution: Some individuals experience allergic reactions to plantain.

PUFFBALLS
Lycoperdon and *Calvatia* spp.

Puffballs are a distinctive fungus, round to pear-shaped in appearance. Though they lack a true stem, they have a stalklike base. They release spores through a hole or cracks in the surface. Size varies from marble to baseball and larger.

Derivation of name: *Lycoperdon* translates as "wolf's fart." *Calvatia* means "bald."

Other name: devil's snuffbox.

Range: throughout Alaska.

Harvesting directions: Cut each puffball in half and examine carefully. Discard any that reveal the outline of a mushroom cap and stem inside, as this may be the button stage of the deadly *Amanita*. Specimens suitable for consumption are creamy white, and homogeneous inside; flesh should look like smooth cream cheese (see center of photo). Avoid eating any with yellow, discolored, or mushy flesh or with black or jellylike insides (see top of photo).

Food use: Harriette Parker, President of the Alaska Mycological Society, recommends peeling large puffballs before cooking. Large puffballs can be sliced and cooked as steak and served topped with puffball gravy. Add puffballs to soups and stir-fries.

Medicinal use: The tumor-inhibiting drug Calvatin is extracted from *Calvatia* puffballs and used in cancer research.

Other: Puffballs thrive in lawns and in hard-packed gravel of abandoned roads. Though stepping on old puffballs and watching them "puff" spores is a favorite children's activity, I discourage the practice as inhaling fungal spores can irritate bronchial linings and cause allergies.

Caution: Follow harvesting directions above carefully. A single mushroom releases millions of spores.

SHEPHERD'S PURSE

Capsella bursa-pastoris   Mustard family (Brassicaceae)

Shepherd's purse is easily recognized by its distinctive heart-shaped seedpods. The small flowers have four white petals and six yellow stamens. Upper stem leaves are arrow-shaped and alternate along the stem; the lower leaves, like those of dandelions, are deeply lobed and arranged in a basal rosette.

Derivation of name: *Capsella* means "little box"; *bursa-pastoris* translates as "shepherd's purse."

Other names: lady's purse, pickpocket, mother's heart, St. James' weed, toywort, salt and pepper, poor man's pharmacetty.

Range: Southeast Alaska to Kodiak and the western Yukon and north to the Brooks Range.

Harvesting directions: Leaves are prime before flowering; flavor becomes stronger and more peppery with age. Pick flowers and seedpods throughout the summer.

Food use: Add leaves to salads, stir-fries, and soups. Seeds and roots are traditional spices.

Medicinal use: Shepherd's purse is high in calcium, potassium, sulfur, vitamin C, and the blood-clotting vitamin K. The mashed herb is a traditional poultice for cuts. The tea is drunk to soothe stomach ulcers and as a remedy for internal bleeding. Add leaves to postnatal baths. For a nosebleed, place crushed leaves in the nostril. The herb dries poorly; for long-term use it is generally tinctured. German herbalist Maria Treban advises rubbing the tincture on hernias.

Other: Shepherd's purse was introduced to the Americas by the Pilgrims. The Chinese cultivated the greens for food and medicine. University studies indicate that seeds, placed in stagnant water, release a gummy substance that destroys mosquito larvae.

WORMWOOD
Artemisia tilesii   Aster family (Asteraceae)

Wormwood has irregular lobed leaves that are silvery underneath. Crush a leaf and you'll notice its distinctive aroma. Though Natives call it stinkweed, I find the aroma quite pleasant. Flowers are non-descript balls, yellow-green to brown in color. Stems generally grow 1 to 3 feet in height, but I've encountered exceptional specimens reaching 6 feet.

Derivation of name: *Artemisia* honors Artemis, Greek goddess of the moon and female energies.

Other names: stinkweed, caribou leaves, Alaskan sage.

Range: northern Panhandle to the Bering Sea and Arctic Ocean.

Harvesting directions: For food use, gather leaves before flowering. For medicinal use, both leaves and flowers can be harvested.

Food use: Wormwood is highly bitter, and food use is only as a spice. Add small amounts of the dried, ground herb to bread stuffings. It flavors New Mexican dishes and cornmeal.

Medicinal use: Wormwood is probably Alaska's most widely used medicinal herb. Natives statewide apply it externally in hot packs for sore muscles, skin tumors, and infections. Wormwood teas treat colds, flu, and upset stomach, though I prefer the tincture—taken in dropper doses—at the first sign of a cold.

Other: Add wormwood to liniments and massage oils; *Artemisia* is a major ingredient in a commercial liniment used by Alaskan fishermen to bathe hands and prevent fish poisoning. Historically, Europeans consecrated wormwood in bonfires on the summer solstice and hung the herb in pastoral homes for protection from evil. To fight fatigue when hiking, travelers placed wormwood leaves in their shoes.

Caution: Use wormwood internally in small quantities only. The volatile oil absinthol, present in large quantity in the California species *A. absinthium*, also exists in variable degree in other *Artemisia* species. Large doses of absinthol can cause coma and convulsion.

BIRCH

Betula spp.   Birch family (Betulaceae)

photo by Norma Wolf Dudiak

Birches range worldwide throughout the temperate regions and vary widely in size, from low-growing bog shrubs to towering trees with whitish or grayish bark. Deciduous leaves turn gold in autumn. Flowers are drooping catkins. Birch stems have a sandpaper texture.

Derivation of name: Birch may be derived from the Sanscrit word *bhurja*, a tree whose bark is used for writing.

Other names: lady birch, lady of the forest, paper birch, Kenai birch.

Range: from Kodiak to the Brooks Range.

Harvesting directions: Collect sap in very early spring by drilling a ½-inch-diameter hole (at a slight upward angle) on the sunny side of the tree. Insert a sap tap (whittle one from an elder stem, removing all the pith) and hang a bucket to collect the sap. Plug the hole with sphagnum moss when done. Collect leaves when bright green and the size of your fingernail.

Food use: Drink the sap as a refreshing tonic brew. Freeze sap for winter use. For pancake syrup, boil the sap (preferably outdoors) until a dark, sweet concentrate results; depending on sugar content, it takes 80 to 100 gallons of sap to yield 1 gallon of syrup. Reduce 10 gallons of sap to 1 gallon for making vinegar or beer. Steep young twigs in hot sap for a spicy drink. Add spring leaves to salads.

Medicinal use: Gargle with fresh sap for mouth sores. Add buds and leaves to skin salves for ringworm. Bark decoctions are used internally for fevers and diarrhea and externally for skin afflictions. Leaf infusions are said to ease urinary problems and kidney stones.

Other: Sip birch sap as a refreshing sauna drink; switch yourself with a leafy birch branch to stimulate your skin.

Caution: Due to salicylic acid content, birch decoctions can cause problems for those hypersensitive to aspirin.

BLUEBERRY

Vaccinium spp.   Heath family (Ericaceae)

Blueberries thrive in the temperate zone worldwide. Alaskan species vary from dwarf bog specimens to 5-foot shrubs. Flowers are attractive bells, pink to white in color. Red-fruited species like *V. parvifolium* (often called red huckleberries) have a smaller distribution and recognition factor.

Derivation of name: *Vaccinium* is the classical name for blueberry and cranberry.

Other names: huckleberry, great bilberry, whortleberry, dyeberry, wineberry, Mother's Day flowers.

Range: throughout Alaska except the extreme north Arctic.

Harvesting directions: The early-blooming flowers are edible and sweet, but most gatherers prefer waiting for the delectable fruits!

Food use: Blueberry pies, cobblers, pancakes, jams, and ice cream are part and parcel of the American cuisine. Anyone who has ever picked them also knows the delights of snacking fresh from the shrub. Berries freeze and can well. Brew blueberry wine or liqueur. Cook berries to a concentrate for juice or pancake topping.

Medicinal use: Blueberries are high in iron and mineral salts. Fruits are suitable for stimulating the appetite of convalescents. Gargle with the juice for sore throat and gums. Blueberry-leaf tea (collected before fruits ripen) is used by herbalists to help in stabilizing blood sugar. Infusions of the antiseptic leaves treat urinary disorders.

Other: Blueberries are a fine source of natural dye for fabric as well as lip balms. Fruits can be used as fish bait.

Caution: *Vaccinium*-leaf tea, in large doses, can cause nausea and vomiting.

CHIMING BELLS

Mertensia paniculata   Borage family (Boraginaceae)

Chiming bells is related to the familiar garden herbs comfrey and borage. Chiming bells grows knee-high and bears rough, hairy leaves and blue bell-shaped flowers. Like fireweed, it germinates quickly following fire, softening the charred landscape with carpets of color.

Derivation of name: *Mertensia* honors the 19th-century German botanist Franz Karl Mertens; *paniculata* indicates that the flowers are arranged in panicles.

Other names: bluebells, mountain bluebells, lungwort.

Range: Southcentral Alaska west to Bethel and north to the Brooks Range.

Harvesting directions: Leaves are prime before flowering. Pick flowers when fully open, before petals fade.

Food use: The flowers are a trail snack, a garnish, and a good addition to floral gelatin and green salads. Add dry flowers and leaves to beverage teas. The hairy leaf texture inhibits most foragers from adding leaves to salads, but in cooked dishes like casseroles, chiming bells help stretch vegetable supplies.

Medicinal use: A European *Mertensia* species, dubbed lungwort, has reportedly been used for relieving asthma, bronchitis, and other lung complaints.

Other: As a flower essence (see Herbal Directory), chiming bells are said to assist individuals in finding joy in physical existence and peace through understanding one's true nature. I delight in growing chiming bells in the wildflower garden and adding these spirit-brightening blossoms to potpourris.

COTTONWOOD
Populus balsamifera   Willow family (Salicaceae)

Cottonwood is a distinctive tree of gravel bars and tundra. Its thick, gray-brown bark becomes deeply furrowed with age. The fat, sticky leaf buds form in fall (see photo above); in spring, the buds perfume the outdoors as they open. The leaves are smooth, dark green above and paler underneath; in fall they turn a glorious gold.

Derivation of name: *Populus balsamifera* translates as "balsam-bearing poplar."

Other names: balsam poplar, balm of Gilead.

Range: from the northern Panhandle to the Arctic.

Harvesting directions: Harvest the resinous buds for salve-making from fall through spring breakup. Pick catkins in spring. Harvest bark from pruned branches, preferably when the spring sap is flowing.

Food use: The vitamin C–rich catkins can be nibbled raw or added to stews, although their taste is unexciting. Dry ground inner bark extends flour supplies.

Medicinal use: Steep cottonwood buds in almond oil in the top of a double boiler; the strained oil, thickened with beeswax, yields an exquisite "balm of Gilead" salve. Use freely on rashes, cuts, and piles. Bud and bark liniments are applied hot for muscle aches and sprains. Teas and tinctures treat headache.

Other: Cottonwood buds are burned to clear the psyche for dreaming; add dry buds to dream balms and dream pillows. Steep buds in water as a room freshener. I like to force winter branches in a vase in April, perfuming the house with the scent of spring. Athapaskans use cottonwood for sun goggles and fishing floats.

Caution: Like willow and birch, cottonwood contains salicin. Use with caution internally if hypersensitive to aspirin.

CURRANT

Ribes spp.   Gooseberry family (Grossulariaceae)

"Gooseberries" is the common name for prickly-fruited species, whereas "currants" are smooth. Growth habits vary from trailing to upright; many thrive around old stumps. The nicknames stink currant and skunk currant refer to the pungency of the maplelike lobed leaves. Though the red currant is the classic favorite, all species bear palatable fruits and are relished by hungry campers.

Derivation of name: *Ribes* is from the Arabic name for a plant with sour sap.

Other names: gooseberry, dog bramble, black currant, red currant, stink currant, skunk currant.

Range: Southeast Alaska to the northern Alaska Peninsula and northward to the Brooks Range.

Harvesting directions: For preserves, gather fruits when plump but before they are fully ripe, as they are higher in pectin. For snacking and baking, wait until full maturity. For tea-making, gather leaves before flowering, or young, new growth afterward.

Food use: Use currants in all the basic berry dishes: pies, cakes, cobblers. They can be blended with blueberries and other sweeter fruits if desired. Currant-mint jelly is a favorite with lamb. Use leaves fresh or fully dried in beverage teas. Currants are high in vitamin C and copper.

Medicinal use: Red-currant infusions are Dena'ina Athapaskan medicine for colds and flu. Whole currants are said to stimulate the appetite and regulate the bowels. Moscow pharmaceutical studies indicate that the acids in unripe currants promote longevity by preventing body cells from disintegrating.

Caution: Do not use wilted leaves as they can be somewhat toxic.

DEVIL'S CLUB
Echinopanax horridum, aka *Oplopanax horridum*
Ginseng family (Araliaceae)

photo by Jon R. Nickles

Devil's club is a fierce-looking plant, often despised by hikers, though revered by herbalists for its ginsenglike properties. Sharp prickles are borne on woody stems, 6 to 8 feet in height, and on the undersides of lobed dinner-plate-sized leaves. The clusters of small white flowers mature as scarlet red fruits, which form a striking contrast to autumn's golden leaves.

Derivation of name: *Echino* is from the Greek for a hedgehog or sea urchin; *panax* refers to ginseng, and originally meant "all-healing" or "panacea"; *horridum* is from the Latin for "prickly."

Other names: Alaskan ginseng, *suxt.*

Range: Southeast to Southcentral Alaska and the Alaska Peninsula.

Harvesting directions: Pick leaf shoots when they first emerge in early spring; spines must be *soft* to the touch and closed tightly. Dig roots in spring as soon as the ground is workable. To de-spine stems, scrape with a knife or singe in a fire. Chip inner and outer bark of roots and stems for tea.

Food use: The tender soft leaf shoot is a trail snack. Add one or two to soups as a spice. I especially like them in egg-based dishes.

Medicinal use: Devil's club is a tonic tea, drunk by Southeast Alaska Indians as a cancer preventive. Teas and tinctures are said to help stabilize blood sugar of borderline diabetics. The hot, pounded root is applied as a poultice for festering wounds and insect stings.

Other: To relieve cold feet or headache, try a stimulating footbath of boiled devil's-club root. Add root to massage oils and liniments.

Caution: Medical supervision is essential for internal use by diabetics on injectable insulin.

FIDDLEHEAD FERNS
Matteuccia struthiopteris, Dryopteris dilatata, Athyrium filix-femina
Lady fern and shield fern families (Athyriaceae and Aspleniaceae)

Fiddleheads, or croziers, are the coiled edible spring growth of ferns. Ostrich fiddleheads are smooth at emergence, whereas lady and wood ferns bear a brown woody chaff. Mature wood fern (also called shield fern or trailing wood fern) has a triangular blade; both ostrich and lady fern are plumelike, tapering at the base and top. Lady and wood ferns bear spores on the undersides of the green leaves. Ostrich ferns have a central shorter spore-bearing frond.

Derivation of name: *Struthiopteris* means "ostrich feather"; *dryopteris* translates as "oak feather"; and *filix-femina* means "lady fern."
Range: from the Brooks Range southward to the Aleutian Islands and the Alaska Panhandle.
Harvesting directions: Pick the fiddleheads (2 or 3 from each cluster) when tightly coiled in early spring. Remove chaff from lady and wood ferns. In late fall, fiddleheads can also be found just below the surface of the ground. Dig underground rootstalks in early spring and in fall after frost.
Food use: Sauté fiddleheads in butter or cook as a potherb and top with garlic or your favorite sauce. Even finicky children love batter-dipped fried fiddleheads. Freeze or can fiddleheads for year-round use, or make fiddlehead pickles. Fiddleheads are high in iron, potassium, and vitamins A, B, and C. The boiled, peeled roots are a good emergency food source.
Medicinal use: Shield-fern rhizomes are a traditional vermifuge.
Caution: Mature fern fronds are toxic. Collect fiddleheads only when fully coiled. Consumption of bracken fern (*Pteridium*), a species that ranges in Southeast Alaska, has been linked to stomach cancer.

HIGHBUSH CRANBERRY
Viburnum edule   Honeysuckle family (Caprifoliaceae)

photo by Norma Wolf Dudiak

Taste a raw highbush cranberry, and you're apt to grimace. But sample highbush cranberry cake or vibrant highbush jelly, and you're likely to grin. Observe the floral parts, which are in an opposite arrangement. The maplelike leaves are lobed except for the uppermost pair. Flowers are in flat clusters, with individual blossoms bearing five petals and five stamens. The tart fruits each bear one large flat stone.

Derivation of name: *Viburnum* is Latin for "wayfaring tree"; *edule* means "edible."

Other names: crampbark, mooseberry, highbush berry, squashberry, sheepberry.

Range: Southeast to the Alaska Peninsula and north to the Arctic; rare in westernmost and northernmost Alaska.

Harvesting directions: Many foragers gather highbush fruits before frost, when they are higher in pectin and have a fresh aroma. Some prefer harvesting after cold snaps, as chilling sweetens the fruits somewhat. Flowers can be picked in early summer. Harvest bark from pruned stems in spring or fall.

Food use: The only raw highbush cranberries I truly savor are those found on winter expeditions; the frozen fruits provide a refreshing natural sherbet for thirsty skiers. Cooked fruits, run through a food mill, make tasty jellies, catsup, syrups, and sauce. Flowers can be added to pancake batters.

Medicinal use: Bark decoctions are sipped to relieve stomach and menstrual cramps. The muscle-relaxing properties of "crampbark" are due to the bitter glucoside viburnine. Alaskan Natives use bark decoctions externally for infected cuts.

Other: British Columbia Indians stored steamed, slightly underripe fruits in water in cedar boxes. Patch ownership was highly esteemed.

SALMONBERRY
Rubus spectabilis   Rose family (Rosaceae)

photo by Jon R. Nickles

Salmonberry canes, which bear weak spines, soar to 7 feet and often form dense thickets. The bright reddish-purple, early-blooming flowers are a hummingbird favorite. The raspberrylike fruits are thumb-sized, and are red or orange-gold in color at maturity.

Derivation of name: *Rubus* means "bramble"; *spectabilis* translates as "exceptionally showy."

Other name: muck-a-muck.

Range: primarily a coastal species, ranging from Southeast to Southcentral Alaska and the Aleutian Islands.

Harvesting directions: Pick salmonberry buds and blossoms as they appear in mid- to late spring. Harvest fruits in midsummer when plump and fully ripe.

Food use: Add buds and blossoms to spring salads. Snack on fresh fruits. Blend salmonberries with yogurt or breakfast cereal. Make salmonberry jams, pies, and herbal vinegars. Leaves and berries yield a pleasant beverage tea.

Medicinal use: Leaf infusions and bark and root decoctions are herbal treatments for diarrhea and dysentery. Explorers drank the brew to settle intestinal upset caused by overindulging in rich salmon. Pounded bark poultices were applied to aching teeth.

Other: Use the astringent leaves and root decoctions externally for oily skin and hair. In Canadian Indian cultures, salmonberry patches were an item of prestige, and picking rights were highly prized.

Caution: Use only fresh leaves or fully dried ones; wilted leaves are mildly toxic.

SPRUCE

Picea spp.   Pine family (Pinaceae)

photo by Norma Wolf Dudiak

Both black and white spruce have four-sided needles that roll easily between the fingertips; black spruce has dense orange fuzz on its new growth stems, whereas white has hairless twigs. Sitka spruce has needles that are flattened and resist rolling.

Derivation of name: *Picea* is the classical Latin name for spruce.
Other names: white spruce *(P. glauca)*, Sitka spruce *(P. sitchensis)*, black spruce *(P. mariana)*, Lutz spruce (hybrid).
Range: Southeast Alaska to the Brooks Range.
Harvesting directions: Gather the bright green spring growth at the end of spruce branches. Gather inner bark from branch cuttings in early spring when the sap is flowing.
Food use: Place a generous handful of spruce tips in a teapot; add boiling water and steep gently 5 minutes. Strain, and serve as a relaxing beverage high in vitamin C. If desired, add orange or lemon slices, cinnamon sticks, cloves, and honey. Inner bark is a traditional survival food, eaten raw or boiled, or dried and ground into flour.
Medicinal use: Captain Cook rationed spruce beer to his crew, keeping spirits up and bodies free of scurvy. Boil needles and inhale steam for sinus infection. Spruce jelly and syrup soothe sore throats. Spruce sap and melted pitch are used for medicinal plasters and protect wounds from infection.
Other: Grind inner bark for a foot powder. Spruce is a legendary "partner plant" of the Dena'ina. Koyukon Athapaskans believe white spruce has a potent and kindly spirit.
Caution: Essential oil of spruce (obtained by distillation) can irritate the skin if applied undiluted.

TWISTED STALK
Streptopus amplexifolius   Lily family (Liliaceae)

The botanical name aptly describes the twisting stems, which grow to 3 feet, and the clasping leaves, which bear parallel veins. Flowers are greenish-white, with petals that curve back. Each red fruit hangs singly on a kinked stem. (See Caution, following.)

Derivation of name: *Streptopus amplexifolius* means "the twisted stalk with the clasping leaf."

Other names: watermelon berry, wild cucumber, scoot berry.

Range: Southeast Alaska to the Aleutian Islands and northward to the Yukon River.

Harvesting directions: In early spring, pick the tender above-ground portion (stem and leaves). Harvest the watery fruits in summer when plump and fully red.

Food use: The young cucumber-flavored shoots are a delicious trail snack or addition to tossed salads. Fried as tempura, the flavor becomes quite similar to asparagus. The ripe fruits taste like watermelon. They can be used fresh as a snack, mixed with more plentiful berries in desserts, or dried and added to cookies.

Medicinal use: In large quantities, the berries can be laxative—hence the nickname "scoot berry."

Caution: Be vigilant when harvesting, as the deadly false hellebore (*Veratrum* spp.) often grows nearby (see Poisonous Plants). Hellebore stems are stouter and bear parallel-veined leaves with deep pleats. A nibble can cause constriction in the throat and breathing difficulty. In the Turnagain Arm–Anchorage vicinity and in the southern Panhandle, the similar-looking but bitter-tasting false Solomon's seal (*Smilacina* spp.) occurs. Raw false Solomon's seal is purgative, hence its poisonous reputation. Botanist Verna Pratt points out that Solomon's seal lacks the black hairs on the lower stem that are characteristic of the choice edible twisted stalk.

VIOLET

Viola spp.   Violet family (Violaceae)

Violets, like roses, are well recognized worldwide, having been companions of humankind in woodland and garden for eons. In optimum conditions, the dainty flowers can reach 10 inches in height. The five irregular petals can vary from traditional violet hues to snow-white and bright yellow. Blossoms are borne on small heart-shaped leaves, and the eventual seeds are in three-pronged capsules.

Derivation of name: *Viola* is said to be the Latin name for the Greek nymph Io. After Io was transformed into a cow, Zeus caused her tears to become violets.

Other names: wild violet, Alaska violet, marsh violet, yellow violet.

Range: throughout Alaska, except the extreme north Arctic.

Harvesting directions: Pick leaves and flowers throughout the green season. Though leaves are tenderer before bloom, they stay mild in flavor.

Food use: Two average-sized violet leaves are said to supply your daily requirement of vitamin C. Add leaves and flowers to green and gelatin salads. Steam leaves as a potherb, or add to soups, omelettes, stir-fries, and casseroles. Use flowers in vinegar. Try violet wine, which was a favorite in ancient Rome. Candy the flowers as a cake decoration.

Medicinal use: Dena'ina Athapaskans burn roots of marsh violet (*V. epipsela*) to ward off disease. Use leaves in skin salves for cuts and scrapes. Poultice the herb for inflamed bruises. Sip violet syrup or floral tea to ease constipation.

Other: Violet lotions soften and perfume the skin. Violet blossoms, infused in fresh goat's milk, make a famous Celtic beauty wash for the skin. Violet toilet water beautifies the complexion.

Caution: All violets, eaten in quantity, can be somewhat laxative, but yellow-flowered species are said to be especially so.

COLTSFOOT
Petasites spp.   Aster family (Asteraceae)

Coltsfoot is highly adaptable, growing in diverse habitats from moist stream beds and tundra to woodland and alpine passes. Leaves vary from triangular to lobed and bear a thick felty covering on their undersides. Flowers frequently appear before leaves develop, a habit that gives rise to one of the plant's nicknames, son before father. The fragrant cluster of blossoms sits atop a rather thick and hairy stem.

Derivation of name: *Petasites* translates as "broad-brimmed hat" and refers, rather imaginatively, to the shape of the leaves.

Other names: sweet coltsfoot, butterbur, coughwort, British tobacco, flower velure, pestilence-wort, son before father, owl's blanket, wolverine's foot, penicillin plant.

Range: throughout Alaska.

Harvesting directions: Pick coltsfoot flowers in early spring, before they start to turn brown. Leaves, for food use, are best when young and tender. Rootstalks are dug spring or fall.

Food use: Steam, sauté or tempura coltsfoot flowers as a side dish. A few young leaves can be chopped and added to casseroles, herb pies, and other dishes where their felty texture won't be noticeable. Roasted roots are eaten by the Siberian Eskimos.

Medicinal use: For soothing a cough or sore throat, I blend a syrup of coltsfoot, licorice root, cherry bark, and slippery elm. The antispasmodic petasin is derived from *Petasites* species. Many herbalists favor coltsfoot teas and tinctures for asthma, croup, and menstrual cramps. In Iliamna, local Natives chew the raw rootstalk of "penicillin plant" for sore throat.

Other: Scandinavians often plant coltsfoot near beehives.

Caution: Large amounts of coltsfoot can cause abortion. Coltsfoot contains pyrrolizidine alkaloids that can irritate the liver, and should not be used for extended periods or in high doses. Herbalist Kathi Keville reports that Germany proposes restricting coltsfoot use to 1 teaspoon dry herb daily for a maximum of 1 month.

JEWELWEED

Impatiens noli-tangere   Touch-me-not family (Balsaminaceae)

Jewelweed is an ideal flower for children's gardens, but youngsters of all ages delight in touching the mature pods to catapult the seeds. Jewelweed thrives in moist shady areas. The slipper-shaped yellow flowers taper to a narrow end with a curved spur. Coarsely toothed leaves alternate along the smooth stems.

Derivation of name: *Impatiens* means "impatient"; *noli-tangere* translates as "no-touch." Both refer to the seedpods that pop open if you touch them.

Other names: touch-me-not, slipperweed, quick-in-the-hand, snapweed.

Range: Southeast and Southcentral Alaska, and sporadically through Western Alaska and the Interior.

Harvesting directions: Pick young shoots and leaves when under 1 foot tall. Gather flowers from tall summer stalks. Collect seeds when pods spring at the touch.

Food use: Young greens are suitable as a vegetable, provided they are cooked. Boil in two changes of water for ten minutes each time, and discard cooking fluids after use. Seeds can be nibbled raw, or sprinkled as a poppy seed replacement on biscuits and cakes. Flowers are an edible garnish.

Medicinal use: Rub crushed jewelweed on the skin to relieve itching of mosquito bites, or sores from poison ivy or cow parsnip. You also can extract the juice for this purpose with a juice machine. An alternative treatment is washing with a decoction of stems and leaves. Add the fresh herb to salves for athlete's foot, warts, ringworm, and skin rashes.

Other: Seeds are good feed for wild or caged birds. Some species have been used as dyes to color fingernails.

Caution: Avoid eating raw older plants; they can be purgative. Eat jewelweed in moderation only.

MONKEYFLOWER

Mimulus guttatus   Figwort family (Scrophulariaceae)

photo by Norma Wolf Dudiak

Monkeyflowers roam the globe, from Alaska to Mexico and as far south as New Zealand. They are readily recognized by their cheery yellow blossoms that have a tubelike structure, irregular petals with a long lower lip, and red spots inside the flower. Leaves have irregular teeth and are in an opposite arrangement, with the lower leaves having a short stalk, whereas the upper are sessile (lacking a stalk).

Derivation of name: *Mimulus* means "mimic"; *guttatus* means "speckled."

Other names: yellow monkeyflower, common monkeyflower.

Range: Southeast and Southcentral Alaska to the Aleutian Islands and north to the Yukon River.

Harvesting directions: Though monkeyflowers are often collected before flowering, they remain palatable throughout the green season. Pick blossoms at their peak. Greens can sometimes be found in winter, under the ice.

Food use: "Greek creek salad" is one of my favorite camping meals. Toss monkeyflower greens, spring beauty, saxifrage, and other suitable wild edibles together and top with feta cheese and olives. Garnish with monkeyflower blossoms and add a touch of dressing. Flowers are also a good addition to homemade gelatin salads and floral salads.

Medicinal use: Stems and flowers can be mashed and applied as a poultice for insect bites and minor cuts and scratches.

Other: Monkeyflowers were a traditional potherb of the American settlers. The plant grows enthusiastically in gardens.

MOUNTAIN SORREL
Oxyria digyna   Buckwheat family (Polygonaceae)

"The pause that refreshes" is a commercial slogan that could easily apply to mountain sorrel. When hiking, I always enjoy nibbling the tart, lemon-fresh leaves. The ground-hugging plant has a compact basal cluster of leaves. Each round to kidney-shaped leaf is on a single stem. The flowering stems grow 6 to 12 inches high and bear reddish-green flowers followed by clusters of flat seeds.

Derivation of name: *Oxyria* refers to the sharp, pungent taste of the leaf; *digyna* refers to its two floral carpels.

Other names: sorrel, sourgrass.

Range: throughout Alaska.

Harvesting directions: Gather green sorrel leaves in spring and summer. Avoid the discolored red leaves of older plants.

Food use: The tart lemony leaves of sorrel are a pleasant refreshment on camping trips. Add leaves to salads, salad dressings, sandwiches, casseroles, soups, and other campfire creations. Sorrel's tangy flavor blends especially well with fish. Blend sorrel with butter for a delicious spread. Simmer leaves in water to make "lemonade" and spike with honey or sugar.

Medicinal use: Sorrel leaves have been used as a poultice for warts and skin irritations.

Other: Both wild and cultivated sorrels are easy to grow in northern gardens. The closely related *Rumex acetosa* is the base of a famous French *soupe aux herbes*. Look for *Oxyria* along abandoned roads and in moist mountain meadows.

Caution: Consume sorrel in moderation only; the leaves contain oxalic acid, which binds with calcium in the body, forming calcium oxalate crystals that can damage the kidneys.

NAGOONBERRY

Rubus arcticus   Rose family (Rosaceae)

Nagoonberry is eye-high to a mouse but of interest to far more than rodents. The flowers are an eye-catching hot pink. Leaves may be partially or fully divided into three lobes. Fruits are red, made up of many drupelets, resembling small raspberries (to which they are related).

Derivation of name: *Rubus arcticus* translates as "Arctic bramble."
Other names: dewberry, wineberry, arctic raspberry, bramble dewberry.
Range: Southeast Alaska to the Aleutians and north to the Brooks Range.
Harvesting directions: Pick the flowers at their peak, before petals begin to wither. Gather fruits when sweet and soft. Leaves are prime before flowering.
Food use: Nagoonberry blossoms are a sweet trail nibble. Add fruits and leaves to beverage teas. Nagoonberries are delectable but difficult to gather in quantity. Inupiat Eskimos use *Rubus arcticus* to heighten flavor of the more abundant *R. chamaemorus*.
Medicinal use: Leaves and roots of various *Rubus* species are used for relieving diarrhea and dysentery. Nagoonberry fruits are extremely high in vitamin C and would be useful in treating scurvy.
Other: The Dena'ina Athapaskan name translates as "frog's berry" or "frog's cloudberry." Nagoonberries thrive in bogs, moist meadows, and along stream banks. In the early 1970s, the berries were marketed by an Alaskan preserve company for $45 pint.
Caution: Use leaves fresh or fully dried. Wilted leaves can cause nausea and vomiting.

SAXIFRAGE

Saxifraga spp.   Saxifrage family (Saxifragaceae)

The saxifrage genus is large and can be somewhat confusing, but the good news is that all species are edible. My preferred species for eating is brook saxifrage, *S. punctata*, which has smooth, kidney-shaped leaves with toothed margins. Leaves grow singly on basal stems. White flowers are borne in a spike; look closely and you'll note ten stamens with a very visible ovary in the center. Seed capsules are red. A plant sometimes mistaken for brook saxifrage is mist maiden (*Romanzoffia*); leaves are similar in shape and equally edible.

Derivation of name: *Saxifraga* is from the Latin *saxum*, "rock," and *frangere*, "break." This herb thrives in cracks in rocks, hence the belief that it breaks rocks. Another theory of derivation relates to saxifrage's ancient use in dissolving urinary stones.

Other names: brook saxifrage, salad greens, deer tongue.

Range: throughout Alaska.

Harvesting directions: Leaves are best before flowers appear.

Food use: Add saxifrage leaves to salads or stir-fries. Use in quiche, spanakopita, and casseroles. Though it is hard to gather in quantity, saxifrage helps stretch supplies of more abundant and tastier greens. Leaves are high in vitamins A and C.

Medicinal use: Current medicinal use of saxifrage appears relatively uncommon except in China, where species are used to treat ear infections and upset stomach. The nutrient-rich greens are suitable for treatment of scurvy and vitamin deficiency.

Other: Saxifrage was a popular salad herb of the Pennsylvania Dutch. Some carpet-forming species, such as *Saxifraga bronchialis*, are used as garden groundcovers.

SPRING BEAUTY

Claytonia siberica and *C. sarmentosa*   Purslane family (Portulacaceae)

The dainty long-flowering spring beauty thrives along creeks and in moist, shady locations. Flowers have five pink or white petals. Leaves are on single stems in a basal cluster and in pairs on the flowering stems. The greens remain mild-flavored throughout the growing season. All species in the *Claytonia* genus are edible.

Derivation of name: *Claytonia* honors 17th-century botanist John Clayton; *siberica* means "of Siberia"; *sarmentosa* refers to runners (stolons).

Other name: Siberian spring beauty.

Range: Southeast and Southcentral Alaska to the Aleutian Islands and the Arctic.

Harvesting directions: Stems are rather stringy, so it's best to pinch off the leaves and blossoms. Be certain to leave an abundance of flowers to set seed and propagate the species.

Food use: Add spring beauty leaves to salads and vegetable dishes. Flowers are an attractive edible garnish on spreads, cakes, and hors d'oeuvres. Leaves are high in vitamins A and C.

Medicinal use: Since greens are high in ascorbic acid, they could be of use in treating scurvy.

Other: Some *Claytonia* species (*C. acutifolia* and *C. tuberosa*) have fleshy taproots or fleshy corms that are used as potato substitutes. Indiscriminate harvesting has endangered the plant in some regions. The annual *C. sibirica* grows well in garden soil and apparently self-seeds freely. The perennial *C. sarmentosa* has fewer flowers and produces bud-bearing runners.

CLOUDBERRY

Rubus chamaemorus   Rose family (Rosaceae)

In northern Alaska you'll find cloudberry called salmonberry, whereas in the south *Rubus chamaemorus* is cloudberry and the tall woody shrub *Rubus spectabilis* is salmonberry. Use whatever common name you prefer, but do become acquainted with this raspberry relative. Cloudberry grows as high as your ankle and bears one five-petaled white flower and one to three lobed leaves per plant.

Derivation of name: *Rubus* means "bramble"; *chamaemorus* translates as "ground mulberry."

Other names: salmonberry, knotberry, baked appleberry, *akpik*, ground mulberry.

Range: throughout Alaska.

Harvesting directions: Pick the fruits when plump and golden-red. Immature fruits are hard and deep red. Flowers are edible, but fruits are generally preferred. Leaves for tea-making are prime before flowering.

Food use: Munch cloudberries raw; these tasty fruits are an excellent source of vitamin C. Combine with crowberries, bog cranberries, and lingonberries and top with yogurt. Blend in fruit smoothies. Process into jam or jelly. If you imbibe alcoholic spirits, try the Scandinavian favorite cloudberry liqueur. Fresh or dry leaves (not wilted) can be added to tea blends. The Inupiat eat cloudberries with oil to add flavor and prevent constipation.

Medicinal use: Drinking seedless cloudberry juice is a Yup'ik remedy for hives.

Other: Alaskan Natives traditionally preserved cloudberries by storing them in seal oil. Cloudberries are also found in bogs.

CROWBERRY
Empetrum nigrum   Heath family (Ericaceae)

Crowberries—or blackberries, as they're commonly called in northern Alaska—have seedy blue-black fruits that grow on trailing stems in an evergreen carpet. The narrow, needlelike leaves have edges that roll back and meet. South of Anchorage, crowberries have separate male- and female-flowered plants, so if you happen upon a patch of "bachelors" you'll find yourself berryless. North, the flowers are bisexual, but you'll need a magnifying glass to appreciate their structure.

Derivation of name: *Empetrum* is from the Greek meaning "upon a rock"; *nigrum* means "black."

Other names: blackberry, mossberry, black crowberry.

Range: throughout Alaska.

Harvesting directions: Gather fruits after fall frosts. If you gather the fruits earlier, place them in the freezer so the cold can sweeten the taste. Pick overwintered fruits after snowmelt.

Food use: Add crowberries to muffins and cakes. Process into jams and jellies. Try making crowberry juice, wine, or liqueur. Freeze, can, or dry crowberries for winter use. Dried crowberries can be used alone or with serviceberries in pemmican recipes. Alaskan Natives preserve crowberries in seal oil. Inupiat Eskimos also blend them with fish livers.

Medicinal use: Dena'ina Athapaskans drink crowberry decoctions for upset stomach and diarrhea. The berry juice and root decoctions are a traditional remedy for sore eyes.

Other: Crowberries are an attractive evergreen for the home terrarium.

LABRADOR TEA
Ledum spp.   Heath family (Ericaceae)

An herb walk I led for a mother and child ended abruptly at the Labrador tea patch. The four-year-old was enchanted by *Ledum's* fragrance and intent on collecting leaves for small pillows to perfume her clothes. This woody-stemmed shrub can be easily identified by its leaves, which bear a brownish felt on the underside; crush the leaf and you'll discover its captivating scent. Flowers are white to pink, borne atop the shrub.

Derivation of name: *Ledum* is derived from the Greek *ledon*, for a plant bearing similar leaves. The specific name *palustre* means "marsh"; *groenlandicum* means "of Greenland."

Other names: Hudson's Bay tea, marsh tea, moth herb, Greenland tea, trapper's tea.

Range: throughout Alaska except for the Aleutians.

Harvesting directions: Though I generally pick green summer leaves and blossoms, some foragers prefer the brown winter leaves for tea. Be certain to check leaves for their distinctive fragrance and felty under-leaf coating. Leaves of toxic bog rosemary are similar in appearance but are white underneath and lack a pronounced odor.

Food use: *Ledum* is a traditional tea, made by pouring boiling water over the leaves and steeping lightly. (See Caution, following.) For a warming toddy on a blustery camping trip, add a nip of brandy or rum. Use as a spice to flavor black tea, marinades, or stews.

Medicinal use: Yup'ik Natives sip boiled Labrador tea for food poisoning or upset stomach. Dena'ina Athapaskans drink *Ledum* infusions for heartburn, tuberculosis, colds, and arthritis. The tea is also recommended for hangover and constipation.

Other: Add leaves to potpourris and sachets to repel moths.

Caution: Individuals with heart problems and high blood pressure should avoid use of Labrador tea. For others, moderation is strongly advised. The herb contains ledol, a narcotic toxin that can cause drowsiness, cramps, and heart palpitations.

LINGONBERRY
Vaccinium vitis-idaea   Heath family (Ericaceae)

photo by Norma Wolf Dudiak

Delectable lingonberries stir memories of hours spent crawling on the Bristol Bay tundra, collecting the winter's supply of berries. The prized fruits thrive on low, creeping shrubs. Look closely at the underside of the oval leaves and you'll notice their characteristic dark dots. Flowers are pinkish bells; the mature red fruits taste like cranberries.

Derivation of name: *Vaccinium* is the classical name for blueberry and cranberry; *vitis-idaea* means "vine of Mount Ida," honoring a mountain in Crete (today called Mount Idhi or Ìdhi Òros).

Other names: lowbush cranberry, cowberry, rock cranberry, mountain cranberry, partridgeberry, foxberry, red berries.

Range: throughout Alaska.

Harvesting directions: Gather lingonberries after fall frost. Overwintered fruits can often be found in spring.

Food use: Use lingonberries for nut breads, liqueurs, and jam. Grind lingonberries with oranges and walnuts, and sweeten with honey or maple syrup, for a delicious holiday sauce. Martha Ellen Anderson of Sterling dries lingonberries, grinds them in a blender, and uses the tangy red powder on Christmas cookies in place of red-dyed sugar. Fresh or dry lingonberries make a delicious tea. For a beverage concentrate that keeps well in the refrigerator, heat lingonberries slowly in a pan with honey and a few drops of water until berries "pop"; use one teaspoon per cup boiling water for a refreshing drink.

Medicinal use: Inupiat Eskimos apply mashed lingonberries as a poultice for sore throat and rashes. Sipping lingonberry juice aids digestion.

Other: Lingonberries have been popular gifts since the 1700s and are equally appreciated today. The fruits contain benzoic acid, a natural preservative responsible for their outstanding keeping qualities.

SERVICEBERRY

Amelanchier spp.   Heath family (Ericaceae)

Serviceberries are well named, as these fruits provide countless services for inhabitants of the North Temperate zone. Fruits are borne on a woody shrub in Alaska, but in milder parts of the country *Amelanchier* can grow to a small tree. Flowers are showy, with five white petals and five sepals. The lower margin of the leaf is smooth, whereas the upper has teeth. The fruits, which botanically are called pomes, are blue-black with many seeds.

Derivation of name: *Amelanchier* is of obscure derivation.

Other names: Juneberry, Pacific serviceberry, serviceberry, Indian pear, shadbush, saskatoon.

Range: sporadically through Southeast Alaska, Southcentral, and the Interior.

Harvesting directions: Pick fruits when plump and blue-black. Gather leaves for tea in spring, or young, new growth later in the season.

Food use: Serviceberries are a good topping on cereal. Try them in pancakes and pies, breads and muffins, crêpes and puddings. Serve raw or stewed. Dry fruits as a raisin substitute. Dried serviceberries are traditionally blended with ground meat and melted animal fat, kneaded into a paste, and packed into sausage casings for a high-energy camping snack. Leaves, as well as berries and shoots, can be steeped as a beverage tea.

Medicinal use: The boiled inner bark was a North American Indian remedy for snowblindness.

Other: The city of Saskatoon, Saskatchewan, Canada, is named after the Indian name for serviceberry. These shrubs were a vital part of Native life, providing food, carving material, and barter. In Klamath Indian tales, the first people were created from saskatoon bushes.

BEDSTRAW

Galium spp.   Madder family (Rubiaceae)

Square stems and leaves arranged in whorls are characteristic of the *Galium*s. The small flowers have four white petals, and vary with species from sparse to dense clusters. Fruits may be paired or singular, bristly or covered with stiff hairs. Cleavers have weak stems and piggyback on neighboring vegetation.

Derivation of name: *Galium* is from the Greek *gala,* "milk." Bedstraw served as rennet for coagulating milk to make cheese.

Other names: goose grass, gravel grass, stick-a-back, cheese rennet, Our Lady's bedstraw (*Galium* spp.), cleavers (*G. aparine*), northern bedstraw (*G. boreale*), sweet-scented bedstraw (*G. triflorum*).

Range: throughout Alaska except for the extreme north Arctic.

Harvesting directions: Collect spring leaves and stems before flowering for food or medicine. Harvest the fruits in late summer.

Food use: The smooth-stemmed bedstraws are preferred for food use; steam lightly to improve texture. Add cooked cooled greens to salads. Seeds can be ground as a coffee substitute. The flowering herb is a substitute for green tea.

Medicinal use: Herbalists advocate cleavers *(G. aparine)* as a tea and a tincture to treat lymphatic-system and urinary-tract imbalances and skin problems. *Galium* juice, ointment, and poultices soothe burns and skin ulcers. Dena'ina Athapaskans apply *G. boreale*, which they call "wormwood's partner," as a hot pack for aches and pains.

Other: The tangled mats of cleavers are beverage strainers for campers, and are used in Scandinavia to filter milk. Roots yield rose madder paint.

Caution: Continued use of *G. aparine* has been known to cause irritation to the mouth and tongue. If used frequently, blend cleavers with a soothing demulcent herb such as bladderwrack.

BURNET
Sanguisorba spp.   Rose family (Rosaceae)

photo by Norma Wolf Dudiak

Young burnet has fanlike leaves in a pinnate arrangement. The fragrant blossoms look like baby bottlebrushes and may be greenish-white (Sitka burnet) or purplish (European burnet). European burnet is renowned as a salad herb, but young Sitka burnet is a palatable substitute. Whether burnet lives up to its reputation of "preserving the body in health and the spirit in vigor" is yet to be proved, but certainly the fresh air and physical exertion involved in hunting for this herb will be conducive to one's well-being.

Derivation of name: *Sanguisorba* is from the Latin *sangis,* "blood," and *sorbeo,* "to stop."

Other names: great burnet, bloodwort, Alaskan bottlebrush.

Range: European burnet, *Sanguisorba officinalis*, ranges primarily from Denali north to the Brooks Range and on the Seward Peninsula; Sitka burnet, *Sanguisorba stipulata,* predominates from Southeast to the Alaska Peninsula and north to Denali.

Harvesting directions: Greens are prime before blooming. For cosmetics and medicinal teas, harvest summer leaves and flowers. Dig roots in fall or early spring.

Food use: Add a few chopped leaves of spring burnet to salads and dressings. Blend with tender, mild-flavored herbs in egg rolls, casseroles, and soups.

Medicinal use: The genus name indicates the herb's astringent properties. Root teas are said to staunch internal and external bleeding. The dry powdered root is used as a styptic for cuts. Flowers and leaves are often added to spring tonic teas.

Other: Burnet is a fine addition to the fragrance garden. It can be propagated by seed or root division. Burnet wine is reputed to cheer the spirits and drive away depression.

CHOCOLATE LILY

Fritillaria camschatcensis   Lily family (Liliaceae)

Giving bouquets of chocolate lilies to unsuspecting recipients is a favorite joke of children. Though these brown lilies are beautiful to behold, the blossoms perfume the air with a manurelike odor. Leaves are arranged in two to three whorls of six leaves, though first-year plants bear a single leaf. Bulbs have white ricelike kernels.

Derivation of name: *Fritillaria* means "dice-box," and is said to refer to the shape of the seed capsule; *camschatcensis* refers to Kamchatka in Russia, where the bulbs were eaten by Natives.

Other names: Indian rice, wild rice, northern rice root, black lily, sarana, Kamchatka fritillary.

Range: Southeast and Southcentral Alaska to the easternmost Aleutian Islands.

Harvesting directions: Dig bulbs in late summer to early fall, when foliage is yellow, indicating that the plant's energy has returned to the bulb, storing sugars and starches. Be certain to return a few ricelike kernels from each bulb to replant chocolate lily. Harvest only in permitted areas and where there is an abundance.

Food use: Bulbs can be bitter and tasteless unless harvested late in the season. I favor the bulbs steamed or stir-fried with other more plentiful vegetables and topped with peanut sauce. Dena'ina Athapaskans soak the bulbs in water before eating. Traditionally, bulbs are dried and ground as a flour extender. They are a good source of starch on wilderness expeditions. The green seedpods are another food source, though I find them rather bitter.

Medicinal use: No medicinal use has been reported for the Alaskan species. In Britain, *F. meleagris* was used as an herb of healing in the 16th century.

Other: The bulbs were a mainstay to Northwest Indians; bulb designs were often carved into cedar boxes and woven into baskets.

COLUMBINE

Aquilegia spp.   Crowfoot family (Ranunculaceae)

Columbines bear showy red or purple flowers with nectar-filled spurs. Stems grow to 3 feet high from a brown carrotlike taproot. There may be three leaflets per basal stem (red columbine) or one leaflet with three deep lobes (purple columbine). Columbine grows easily from seed, which can be collected from the papery capsules.

Derivation of name: *Aquilegia* is from the Latin *aqua,* "water," and *legere,* "collect," and refers to the floral spur tips that contain sweet nectar water.

Other names: red columbine, western columbine (*A. formosa*); blue columbine, dove's foot (*A. brevistyla*).

Range: Southeast and Southcentral Alaska to the eastern Interior.

Harvesting directions: Harvest flowers in summer when fully open. Gather seeds for propagation from the late summer or autumn capsules. Dig roots for external medicinal use in autumn.

Food use: The nectar-filled spurs of columbine flowers are a sweet trail snack. Add flowers to salads as an edible garnish.

Medicinal use: Columbine roots, mashed in olive oil, are an external herbal remedy (recommended by Jeanne Rose in *Herbs and Things*) for relief of aching rheumatic joints. The juice has been rubbed on skin ulcers and boils.

Other: Columbines are a striking addition to your wildflower garden and a favorite of hummingbirds. As an Alaska flower essence, columbine enhances self-appreciation. Omaha Indians used seeds as an aphrodisiac love charm. Hand a bouquet of lovely columbine blossoms to your sweetie, and see what magic unfolds!

Caution: Roots and seeds can cause poisoning if taken internally, due to the presence of a cyanogenic glycoside. Seeds can be fatal to children. Though flowers can be safely eaten, moderation is advised.

COW PARSNIP

Heracleum lanatum   Parsley family (Apiaceae)

photo by Norma Wolf Dudiak

Cow parsnip is of Herculean proportions, growing 6 to 8 feet high. The leaves are the size of dinner plates. Stems are stout, hollow, and woolly-hairy. Note that outer floral petals are longer than the inner petals. Flowers are in umbrellalike clusters (umbels), similar to those of poison hemlock (*Cicuta* spp.). Positive identification is critical (see Poisonous Plants).

Derivation of name: *Heracleum* honors Hercules; *lanatum* means "woolly."

Other names: wild celery, Indian celery, *póochii, pootschki.*

Range: Southeast Alaska to the Aleutians and north to the Yukon River.

Harvesting directions: Use gloves to collect the nonflowering stalks from late spring to early summer. Dig the long, white, parsnip-like roots in early spring or after fall frost.

Food use: Scrape the stems of their woolly covering and wash well. (See Caution, following.) Use peeled, chopped stems as a celery substitute in cream soups, stir-fries, egg rolls, and casseroles. Stems can also be fermented with cabbage for sauerkraut.

Medicinal use: New Mexico herbalist Michael Moore advocates 1 teaspoon dry cow parsnip root in 1 cup water for nausea, acid indigestion, and heartburn. Priscilla Kari reports that Dena'ina Athapaskans use hot roots on toothaches, burn roots as incense during times of illness, and place small bits in meat to worm dogs.

Caution: Harvest cow parsnip with gloves and a knife. Those who harvest bare-handed or weed-whip the plant on sunny days often suffer severe blisters and sunburn. Treatments for these blisters range from jewelweed juice and aloe vera to vinegar and water. Be aware that inhaling the smoke from burning stalks results in internal blisters and can lead to death; children have been severely affected by throwing stalks on campfires and by handling the foliage. Some individuals are allergic to the plant even when it is properly handled.

ELDER

Sambucus racemosa   Honeysuckle family (Caprifoliaceae)

photo by Norma Wolf Dudiak

Elder is the traditional center of herbal gardens and home of the wise grandmother spirit. Tradition dictates that you ask permission before cutting stems or picking flowers; planted near the home, elder brings protection and harmony. The showy clusters of white flowers, sometimes dubbed Alaskan lilac, are followed by attractive red berries that are favorites of birds but bear seeds inedible to humans.

Derivation of name: *Sambucus* is derived from the *sambuke,* a Greek musical instrument made from the hollow woody stem of elder; *racemosa* refers to the type of flower clusters (racemes).

Other names: Alaska lilac, tree of music, red elder, false elder.

Range: Southeast and Southcentral Alaska to the Alaska Peninsula.

Harvesting directions: Gather flowers at their peak in late spring to early summer; flowers pass quickly, so begin gathering when they first appear. Harvest summer fruits only when they are fully red.

Food use: Flowers are my favorite elder product. Add chopped fresh blossoms to pancake, waffle, and cake batters. Fresh flowers, dipped in tempura batter and fried in hot oil until golden brown, are a special treat. Add flowers to herbal tea blends. Use fruits (with seeds removed) for jellies, wines, and beverage concentrates.

Medicinal use: Sip elderflower tea for comfort during a cold; infusions are also traditional for calming the nerves and for alleviating constipation and rheumatic pain. The natural estrogen in the blossoms is said to relieve cramping during menstruation.

Other: Elderflower lotions, toilet waters, and oils are famed for soothing wrinkles and stretch marks.

Caution: Only elder flowers and mature deseeded fruits are safe for consumption. Cyanide poisoning can result from ingesting elder seeds, stems, roots, and immature fruits. Be aware that the toxic baneberry (*Actaea rubra*) also bears red berries but is a small (to 3½ feet) herbaceous plant rather than a tall woody shrub. (See Poisonous Plants.)

FIELD MINT
Mentha arvensis   Mint family (Lamiaceae)

Mint is an excellent herb for sensory awareness. Touch the herb and inhale the refreshing mint fragrance. Feel the square stems and opposite leaves. Observe the opposite, lilac-colored, tubular flowers growing at the junction of leaf and stem.

Derivation of name: *Mentha* is named after the Greek nymph Mintho, who was turned into a mint plant by Pluto's jealous wife, Proserpine. *Arvensis* means "pertaining to cultivated fields."

Other names: wild mint, pole mint, brook mint, Indian mint.

Range: throughout Southeast, Western, and Interior Alaska, with sporadic occurrences elsewhere.

Harvesting directions: Clip in late spring, before flowering, and repeatedly throughout the growing season.

Food use: Enjoy mint tea or cool mint julep as a midday refresher. Add young leaves to spring salads and salad dressings. Leaves are high in vitamins A, C, and K and iron, calcium, and manganese. Serve mint jelly with lamb or wild game. When baking angel food cake, try herbalist Michael Moore's technique. Instead of greasing the cake tin, place mint leaves on the bottom before adding the batter.

Medicinal use: Sip a cup of mint tea if menstruation is delayed or crampy, or if you're suffering minor stomach upset. Try mint for alleviating seasickness or morning sickness. Clear your sinuses by placing your face over a steaming bowl of mint tea and inhaling deeply. Soothe your headache with a cool mint compress.

Other: Splash mint tea on your underarms as a natural deodorant. Add mint to herbal baths and footbaths.

Caution: Mint essential oil is extremely concentrated; an excess can literally burn the skin. Mint can cancel out the effects of homeopathic remedies.

FIREWEED

Epilobium angustifolium   Evening primrose family (Onagraceae)

photo by Norma Wolf Dudiak

Fireweed's showy flowers are four-petaled and vary in hue from bright pink to white. Lower flowers mature first; the blooming of the uppermost blossoms portends the end of summer. Leaves are long, narrow, and willowlike, hence the common name "willow herb." The long pods split and release abundant downy seeds.

Derivation of name: *Epilobium* translates as "upon a pod." *Angustifolium* means "narrow-leaved."

Other names: willow herb, blooming Sally, wild asparagus.

Range: throughout Alaska and the Canadian Yukon.

Harvesting directions: Pick the reddish, asparaguslike early spring shoots of fireweed before the leaves develop; discard the strong-tasting tip (the uppermost inch), and do retain the white, blanched underground portion. (Breaking the underground stem encourages the plant to produce more growth.) Summer leaves are best before blossoms develop. Collect flowers at the peak of perfection.

Food use: Spring fireweed shoots are high in vitamins A and C. Try them steamed as a vegetable, fried as tempura, or pickled with bull kelp. Add a few summer leaves to soups, casseroles, and quiche. The edible blossoms brighten tossed and gelatin salads. Children (and adults) like to split the summer stem and drag it through their teeth to extract the sweet pith. Fireweed flowers yield a beautiful and tasty jelly. For afternoon tea, try a cup of fireweed and mint spiked with delicious fireweed honey.

Medicinal use: Fireweed-leaf tea is used to settle an upset stomach and to gently stimulate the bowels. The boiled herb is a traditional antispasmodic for asthma and coughs. Add leaves, flowers, and the powdered root to salves and boluses for bleeding piles. Root poultices are traditionally used to help draw infection from wounds.

Other: Fireweed regenerates soils after forest fires. The floral essence is said to aid humans in recovering from traumatic experiences.

GERANIUM

Geranium erianthum   Geranium family (Geraniaceae)

Geranium has showy five-petaled purple blossoms with dark purple stripes. Leaves are palmate, with the upper ones sessile (lacking stems) and the lower ones on long stems. Leaves are slightly rough and hairy. Observe the buds and you'll notice downy hairs.

Derivation of name: *Geranium* is from the Greek *geranos,* "crane."
Other names: sticky geranium, cranesbill, stork's bill.
Range: Southeast and Southcentral Alaska to the Aleutians and north to Unalakleet and Denali National Park.
Harvesting directions: Leaves are prime before flowering, but identification is critical. (See Caution, following.) Harvest geranium blossoms when at their peak.
Food use: The edible flowers are a colorful salad garnish and decoration for frosted cakes. I use the young leaves to stretch supplies of delectable greens like goosetongue and lamb's-quarter. Add geranium to egg rolls, casseroles, soups, and spanakopita.
Medicinal use: Powdered roots are astringent and used in traditional treatments for hemorrhoids and intestinal inflammation. New Mexico herbalist Michael Moore advocates the freshly sliced root for gum or tooth infection; apply directly to the painful area. Decoctions of the dry ground root are used for soothing stomach ulcers, diarrhea, and dysentery.
Other: Add geranium leaves and flowers to your bath, and discover for yourself whether the herb is truly an aphrodisiac! Use sticky geranium floral essence for freeing human potential.
Caution: Geranium can be confused with deadly monkshood (*Aconitum delphinifolium*). Both geraniums and monkshood bear palmate leaves, which are divided into five lobed parts (see Poisonous Plants). Geranium leaves are slightly hairy, whereas monkshood leaves are smooth.

GOLDENROD

Solidago spp.   Composite family (Asteraceae)

Goldenrod bears showy clusters of golden blossoms. Stems grow to 2 feet high and have alternate leaves with smaller upper-stem leaves. Leaves vary with species from entire to serrated.

Derivation of name: *Solidago* is said to mean "to make whole."
Other names: woundwort, blue mountain tea, Aaron's rod.
Range: throughout Alaska.
Harvesting directions: Leaves are prime for food before buds form (discard any insect-damaged leaves). Collect flowers for beverage when at their peak in mid- to late summer, stripping flowers from stem. Gather the aboveground herb for medicine just before the peak of bloom. Pick seeds from late August to September.
Food use: Flowers are my favorite goldenrod product, yielding a golden beverage tea. Serve young leaves as a potherb; blend with mixed greens in quiche, soups, and casseroles. Add seeds to stews as a thickener, and flowers to breads and cake batters.
Medicinal use: Goldenrod teas and tinctures are popular herbal remedies for strengthening the kidneys, soothing urinary inflammations and kidney stones, and reducing bronchial congestion. Powdered leaves have been used through the ages for internal and external bleeding. Fresh leaves can be used as a poultice for scrapes and insect bites. Add leaves to skin salves and ointments.
Other: Goldenrod and yarrow are two herbs dubbed "woundwort" that were popular during the Crusades for dressing battle wounds. Early Europeans prized *Solidago* for use in clothing dyes, hair rinses, and medicines. Together with Labrador tea, goldenrod was an American Revolution substitute for highly taxed black tea. Goldenrod is often cursed for causing hay fever, but pollen from *Ambrosia* (ragweed) is the culprit. Flowering stems can be used as a steambath switch.

RASPBERRY

Rubus idaeus   Rose family (Rosaceae)

Wild raspberries thrive on bristly stems, 1½ to 6 feet high. Leaflets have irregular teeth and number three to five. Flowers are white with five petals. The red fruits themselves need little description, as they are similar in taste and shape to our familiar garden raspberry.

Derivation of name: *Rubus* means red or bramble; *idaeus* translates as "of Mount Ida."

Other names: garden raspberrry, European raspberry, American red raspberry, framboise.

Range: Southeast to Southcentral Alaska and north to the Yukon River.

Harvesting directions: Harvest young leaves throughout the growing season. Flowers are edible, but foragers generally wait for the savory summer fruits. Dig roots spring or fall.

Food use: Ways to use raspberry fruits are legion—as snacks or in fruit leather, liqueurs, syrups, jams, and vinegars. Freeze or can raspberries for year-round use. Dry the leaves for tea; add fresh or dried summer fruits for a flavorsome beverage. Fruits are high in vitamins B and C and minerals magnesium, calcium, iron, and phosphorus.

Medicinal use: Raspberry is often a main ingredient in female reproductive herbal formulas. The leaves contain fragrine, which tones reproductive organs. For pregnant women, raspberry tea is reported to aid in decreasing morning sickness and in facilitating birth. My herb students also advocate leaf infusions internally for correcting thyroid imbalances. Rasberry juice and syrup are used to reduce fever in children and adults. Root decoctions treat diarrhea and dysentery. Externally, leaves are a poultice for proud flesh and a wash for wounds. Use infusions as a wash for wounds, mouth sores, and bleeding gums.

Other: Add raspberry leaves to facial steams for oily skin.

Caution: Use leaves fresh or fully dry. Wilted leaves are somewhat toxic.

ROSE

Rosa spp.   Rose family (Rosaceae)

Prickly stems and fragrant five-petaled flowers are characteristic of wild roses. Serrated leaflets are in an odd-pinnate arrangement. Flowers are followed by vitamin C–rich fruits, called hips.

Derivation of name: *Rosa* is from the Greek *rhodon,* meaning "red."

Other names: prickly rose *(R. acicularis)*; Nootka rose *(R. nutkana).*

Range: Southeast and Southcentral Alaska to the Brooks Range.

Harvesting directions: Pick roses in bud or flower stage. Hips are prime after frost.

Food use: Use rose petals (after removing their bitter white base) in salads, teas, jellies, sandwiches, and omelettes. Steep the petals in cool water and spike with lemon and honey for a refreshing julep. Simmer petals in honey in a double boiler for a fragrant spread. Rose hips yield jams, soups, wines, and delectable mock pumpkin pie. Candied hips add flavor to fruitcakes and breads. Hips are high in vitamins A, B, C, E, and K and the minerals calcium and iron.

Medicinal use: Use moistened rose petals as a natural band-aid for minor wounds. Vitamin-rich rose hip tea (three hips contain more vitamin C than an orange) is a traditional drink for cramps, coughs, and colds. Rose hip syrup stimulates the production of red blood cells and is prescribed for anemia.

Other: Simmer rose petals and elder flowers in almond oil in the top of a double boiler for a moisturizing oil for delicate eye areas, stretch marks, and irritated skin. Indulge in rose baths and facial steams.

Caution: The hairs that surround the seeds in rose hips can irritate intestinal linings and cause an "itchy bottom" condition. Straining rose hip tea with a fine filter prevents this. Treat wild birds with a rose hip and popcorn garland.

SHOOTING STAR

Dodecatheon spp.   Primrose family (Primulaceae)

Shooting stars are hardy perennials with exquisite bright-pink flowers; the petals bend back, and the long stamens are said to look like a bird's beak. The basal leaves are light green and taper at the base. The flower stem grows to about 1 foot.

Derivation of name: *Dodecatheon* comes from the Greek *dodeka* and *theoi* and means "twelve gods."
Other names: frigid shooting star *(D. frigidum).*
Range: Southeast Alaska to the Brooks Range.
Harvesting directions: Collect leaves spring through summer. Be careful not to uproot plants after blooming; they may look limp but are storing energy in the roots for the following spring.
Food use: Add leaves to stir-fries, soups, spanakopita, and other cooked dishes. Combine with dandelions in marinated salads.
Other: To start shooting stars for your wildflower garden, Beth Cummings, botanical garden designer at the Pratt Museum in Homer, advises thinly sprinkling seeds on moist sterile soil in a low, well-drained container. Cover seeds lightly with soil and press down. Keep constantly moist. Sprouts generally appear in three weeks and continue appearing for the next two weeks. An alternate method is sprinkling seeds on the soil in a garden nursery box just before freeze-up, and allowing nature to germinate seeds in spring. They bloom in the third season following planting. Gather seed from the wild from late July to September, or see the Herbal Directory for a native seed source.

STRAWBERRY

Fragaria spp.   Rose family (Rosaceae)

Though wild strawberry fruits are smaller than cultivated varieties, their superb flavor makes them a delicacy. White five-petaled flowers bearing five green sepals and yellow anthers are followed by the familar red fruits, whose seeds are embedded near the surface. Leaflets are dark green above with silky hairs below and number three per stalk. Plants send out runners that root at the joint.

Derivation of name: *Fragaria* is from the Latin word for "emit a scent."

Other names: wood strawberry, mountain strawberry, earth mulberry, wild strawberry.

Range: Southcentral Alaska to Kodiak, in the Aleutian Islands, and along the central and eastern Yukon River.

Harvesting directions: Pick fruits when red and juicy. Gather young tender leaves in spring and summer. Flowers are also edible.

Food use: Snack on strawberries, process in jams and syrups, or serve on morning cereal. Strawberry pies, parfaits, ice cream, and wine are other favorites. The leaves, with the addition of mashed fruits, make a tasty tea. Fruits are high in vitamins A and C and minerals calcium, potassium, iron, and sulfur.

Medicinal use: Strawberry leaf tea is an herbal tonic used by pregnant women for alleviating morning sickness, stimulating milk production, and preventing abortion. Whereas the fruits are a gentle laxative, the leaves and roots help tighten loose bowels.

Other: Strawberry-and-cream facials are a beauty aid favored through the centuries by such notables as Marie Antoinette. An old herbal remedy, which has thankfully fallen out of use, blended strawberry juice with the dung of a white dog for throat ulcers.

Caution: Use leaves fresh or fully dried, as leaves are somewhat toxic when wilted. Some people are allergic to the fresh fruits.

WILD CHIVE
Allium schoenoprasum   Lily family (Liliaceae)

Chives are unusual in that the cultivated and wild species are one and the same. Leaves and flowers are edible, but must be differentiated from their lily-family cousin death camas, *Zygadenus elegans*. Whereas chives have a hollow leaf, lilac flowers, and an onion scent, death camas has flat leaves and white flowers, and all parts lack an onion aroma. See Poisonous Plants.

Derivation of name: *Allium* is from the Latin for "garlic" or "onion"; *schoenoprasum* means "reedlike."

Other names: onion grass, garden chive, wild onion.

Range: Southeast Alaska to the Brooks Range.

Harvesting directions: Leaves are prime before flowers appear. If trimmed at the base, they may be harvested repeatedly during the green season. Pick flowers at their peak. Bulbs may be harvested in spring or fall (see identification information above).

Food use: Add fresh or dried chopped chives to dips, spreads, soups, egg dishes, and sour-cream toppings. Chives add flavor to fish and herbal butters. Flowers add zest to salad dressings and seasonings. Chive floral vinegar is a favorite of mine; the striking lilac-colored vinegar with its distinctive onion scent makes superb salad dressings and marinades.

Medicinal use: Chives are said to stimulate digestion. The iron-rich leaves are useful for those suffering from anemia. Crushed chive bulbs were used by Western Indians for insect stings. The bulb juice is used as a natural antiseptic.

Other: Plant chives near peas and lettuce to repel aphids. Rub crushed chives on your skin to discourage mosquitoes.

Caution: Over-mature chive greens can cause digestive upset.

AMERICAN VERONICA

Veronica americana   Figwort family (Scrophulariaceae)

American veronica is an inconspicuous herb of roadside ditches and wet places. The dainty blue flowers resemble those of the Alaskan state emblem, the forget-me-not, but veronica's flowers have four unequal petals and two prominent stamens. Its seed purses are heart-shaped, and its leaves are gently toothed and arranged in pairs.

Derivation of name: *Veronica* is said to honor the saint who wiped Christ's face on the way to the crucifixion.

Other names: speedwell, American brooklime.

Range: Southeast Alaska to the northern Alaska Peninsula, easterly to the Yukon River, and throughout the Aleutian Islands.

Harvesting directions: Harvest these water-loving greens in clean areas; some foragers wash the greens in water with a Halazone tablet added. Though leaves are prime before flowering, I continue to gather the aboveground portions throughout the summer season.

Food use: Japanese and Europeans consider veronica a delicacy and compare it to watercress in flavor. As with dandelions, the greens grow more pronounced in flavor with maturity. Early greens can form the bulk of a salad, later ones a flavor accent. Steam greens as a potherb or add them to stir-fries; they are high in vitamin C.

Medicinal use: Veronica, like coltsfoot, is an expectorant herb, one that helps clear the lungs of mucus. It is a traditional treatment for bronchitis, asthma, and coughs.

Other: According to legend, a shepherd whose king was gravely ill observed an injured deer heal its wound by eating and rolling in veronica. The shepherd reported the sighting; the king regained his health with veronica and showered wealth on the shepherd. Though your acquaintance with veronica may not bring riches, its use can save coins in the household.

BOG CRANBERRY

Oxycoccus microcarpus, aka *Vaccinium microcarpus*
Heath family (Ericaceae)

Bog cranberry is one of the exquisite bog miniatures. Its fine, thread-like stem weaves through the sphagnum carpet, bearing bright pink shooting star–like flowers and plump delicious berries.

Derivation of name: *Oxycoccus* is Greek for "acid berry"; *microcarpus* means "small-fruited."

Other names: bog cranberry, true cranberry, swamp cranberry, moss cranberry.

Range: Southeast Alaska to the Alaska Peninsula and north to the Brooks Range, with sporadic occurrences in the Aleutians and the northernmost Arctic.

Harvesting directions: Pick the berries after frost. You can also pick overwintered fruits in spring, though they're generally sparse and may taste a bit fermented.

Food use: Bog cranberries are tasty raw or cooked. Add to pancakes and nut breads, or grind with rose hips and citrus for a Thanksgiving relish. Process into jams and jellies. Prepare as juice or herbal liqueur. Bog cranberries are high in vitamin C.

Medicinal use: Inupiat Eskimos soak bog cranberries in seal oil and feed to those with poor appetite or gall-bladder difficulties. Yukon River Natives drink cranberry juice for colds and bleeding gums. Cranberry juice is traditionally used as a douche for cystitis and as a drink for urinary-tract infections.

Other: Rub cranberry juice on your skin at night to remove a faded tan, rinsing well each morning and following with a moisturizer. Cranberry is a good source of red dye. Argon dating establishes that this delicate herb has been on the planet for several million years.

CATTAIL
Typha latifolia   Cattail family (Typhaceae)

The sausagelike flowers of cattails progress from green to golden to brown. Cattails can be harvested in all seasons, though exercise caution when gathering, as toxic plants like wild iris also bear sword-like leaves, have rhizomes, and frequent the same habitat.

Derivation of name: *Typha* is the Greek name for "cattail"; *latifolia* means "broad-leaved."

Other names: cat-o'-nine tails, Cossack asparagus, rushes, flags.

Range: sporadically in Southcentral Alaska and the Interior.

Harvesting directions: Pick the female flower in spring when firm and green. The golden pollen from the upper male flower can be gathered by placing a paper bag over the flower, bending the stem, and shaking vigorously. Cut out the cattail heart, the starchy ball at the base of the green stem, and the white rootstalk. In fall, dig the rootstalk. Remove the hornlike sprouts for a vegetable. Process the starch-filled rootstalks for a white flour by scrubbing well and peeling the tough outer rind; then pound with a mallet and cover with water in a jar. The flour will settle to the bottom. Pour off the water, discarding stringy fibers, and use immediately or dry for future use.

Food use: Steam the green female flower briefly as a delicious corn-on-the-cob substitute. Substitute cattail pollen or cattail flour for up to half the flour in biscuit and muffin recipes. Stir-fry cattail hearts.

Medicinal use: Female flowers are added to salves for burns and cuts. The Chinese employ cattail pollen for treating dysentery. Survival teacher Tom Brown advocates rubbing the sticky juice found between cattail leaves on toothaches to dull the pain.

Other: Cattail down is a good filling for children's doll pillows and herbal dream pillows. Leaves are fine caning material.

DOCK

Rumex crispus and *R. arcticus*   Buckwheat family (Polygonaceae)

Docks are adaptable, ranging from the extremes of the Arctic to the tropics. Like dandelions and plantain, they are often dubbed a "weed," but for Alaskan Natives and rural residents they play a large role as food and medicine. The sour-tasting leaves are arranged in a basal cluster; the central stalk grows to 4 feet in height and bears an abundance of seeds in papery reddish capsules.

Derivation of name: *Rumex* is the Latin name for sorrel.

Other names: sourdock, sourgrass, wild spinach; yellow dock, curly dock (*R. crispus*); Arctic dock (*R. arcticus*).

Range: Curly dock ranges sporadically from Southeast and South-central Alaska to the Interior; Arctic dock is more widespread, pre-dominating from Southcentral and Southwest Alaska to the Arctic.

Harvesting directions: Collect leaves for food in spring to summer when bright green in color. Avoid older red leaves. Gather seeds in late summer. Dig the medicinal roots in fall or spring.

Food use: Inupiats blend the boiled herb with berries, blubber, and seal oil and ferment it in wooden kegs. Natives throughout Alaska collect dock extensively and often freeze it for winter use. I add dock greens to casseroles, quiche, salad dressings, and boiled dinners. Seeds can be eaten as a snack or ground as a flour extender. Leaves are high in vitamin C.

Medicinal use: Yellow dock (*R. crispus*) is high in iron; tinctures and decoctions are popular herbal remedies for anemia, hepatitis, liver damage, and skin problems. Yellow dock root is said to help cleanse the system of heavy metals. Add dock root to salves for itching skin. Make a poultice from the root for bee stings and boils. Yup'ik Natives drink dock-leaf juice for colds and upset stomach.

Caution: Consume in moderation only; leaves contain oxalic acid, which depletes calcium in the body.

MARE'S TAIL

Hippuris spp.   Water milfoil family (Haloragaceae)

Though mare's tail and horsetail aren't even remotely related, it's easy to confuse the two. Feel the plant: you'll note that mare's tails are soft in texture and lack the gritty silica of horsetail stems. Tug on mare's tail: it lacks the picture-puzzle capabilities of horsetail, whose stems pull apart and fit back together neatly. Look closely at mare's tail: its leaves are less than an inch long, unlike the growing, jointed leaves of horsetail. (Mare's-tail leaves are arranged in whorls that vary according to species, from four to twelve.) Examine the junction of leaf and stem, where mare's tail's tiny red flowers are located; horsetail's spore-bearing conelike heads are located at the top of the fertile stems.

Derivation of name: *Hippuris* is from the Greek word for "pony."
Other name: goosegrass.
Range: throughout Alaska.
Harvesting directions: Collect the aboveground portion from spring through summer. Natives also gather it in fall after the ice forms and collect brown stems at the end of winter.
Food use: Alaskan Eskimos eat brown, wintered-over mare's tail as an early spring vegetable; the herb changes to bright green when cooked. It is often cooked with goose and duck, as well as boiled in water with seal oil. Yup'ik Natives gather the herb in fall as the ice forms and serve it with fish eggs. I add mare's tail to campfire stews and chowders.
Medicinal use: Mare's tail appears in old herbals as a drink for stomach ulcers, a wash for skin problems, and a poultice for external bleeding.
Other: Cultivate mare's tail in bog gardens to attract wild ducks.
Caution: Gather mare's tail in clean areas. If water quality is in question, foragers can wash mare's tail in water with a Halazone tablet.

MARSH MARIGOLD

Caltha palustris   Crowfoot family (Ranunculaceae)

photo by Norma Wolf Dudiak

Marsh marigold is easily recognized by its round to kidney-shaped leaves and by its bright yellow flowers, which grow on trailing stems that rise up at the ends. Marsh marigold has green seed-filled capsules. Use care in harvesting. See Caution, following.

Derivation of name: *Caltha* is the Latin name for "marigold"; *palustris* means "marsh-loving."

Other names: cowslips, meadowbouts, palsywort, meadowbright, marybuds, horse blobs.

Range: throughout Alaska.

Harvesting directions: Marsh-marigold leaves are one of the first spring greens to appear, and survive the late frosts. Though prime before flowering, greens can be picked throughout the season. Gather flower buds when tightly closed. Dig roots from fall to spring.

Food use: Marsh marigold contains the toxic glucoside proto-anemonin, which dissipates at 180° F. Boil the greens in two to three changes of water. The pickled flower buds are traditionally served as a substitute for capers.

Medicinal use: Marsh marigold is an expectorant and an anti-spasmodic. New Mexico herbalist Michael Moore suggests small doses of a "scant teaspoon or less of the chopped dry herb once or twice a day" for loosening mucus in the chest and sinuses.

Other: Marsh-marigold floral tea, blended with fresh cream, was once used by ladies to beautify their complexions. These lovely herbs flourish in moist places in the home garden.

Caution: Use marsh marigold in moderation only. Daily use can cause kidney or liver inflammation. Handling the raw herb can irritate sensitive skin. Be positive of identification, as toxic wild calla (*Calla palustris*) frequents the same wet habitat; wild calla causes severe burning sensations in the throat if ingested. See Poisonous Plants.

SWEET GALE
Myrica gale   Wax myrtle family (Myricaceae)

Sweet gale is a woody shrub with fragrant foliage. Leaves are smooth on the sides and toothed at the top, hence their descriptive Dena'ina name, "mouse's hand." Crush sweet gale and inhale its sweet perfume. In spring, in areas where the herb is prolific, it can be identified from afar by the golden cast of its pollen-laden male flowers.

Derivation of name: *Myrica* is derived from the Greek "to perfume"; *gale* may be derived from the Greek for "leather helmet," as the male catkins appear in stacked leathery-looking layers.

Other names: gale, Alaskan bay, wax myrtle, meadowfern, piment royal, mouse's hand.

Range: in swampy areas from Southeast to the Alaska Peninsula and through much of western and Interior Alaska.

Harvesting directions: Pick the fragrant buds from fall through spring. Pick leaves in summer.

Food use: Use sweet gale as a substitute for bay leaves. Grind dry leaves and add to herb-salt blends. Buds can be used whole or ground to season stews; as with any spice, a little can add a lot of flavor, so proceed slowly.

Medicinal use: Sweet gale is an old New England remedy for colds and flu. I like the tea blended with cayenne and ginger. Dena'ina Athapaskans drink the brew for tuberculosis. Scandinavians used the boiled herb as an external wash for lice and itchy skin ailments.

Other: Sweet gale switches are well suited for stimulating skin circulation in the sauna. The distilled oil of sweet gale scents male colognes and after-shaves. My attempt at "sweet gale cologne" for my husband ended up being a hit as a hot-toddy base!

Caution: Though safe when used in moderation, sweet gale contains a toxic oil. Vomiting and abortion can result from ingesting the boiled herb. Drink infusions only. Avoid during pregnancy.

WILLOW

Salix spp.   Willow family (Salicaceae)

Moose and humans enjoy eating willow, but both are quite particular about the species they eat. One of the mutual favorites is *surah* (*Salix pulchra*); the long, narrow leaves are smooth on both sides, darker green above, with margins that are generally smooth. The young leaves produce a refreshing aftertaste. The sweet inner bark and peeled shoots of the felt-leaf willow, *Salix alaxensis*, are well liked by Inupiat Eskimos. Though some willows are bitter and generally regarded as inedible, all are safe, so taste can be your guide if you have trouble differentiating among Alaska's three dozen willow species.

Derivation of name: *Salix* is the classical Latin name for willow.
Other names: osier, pussy willow, *surah, churah.*
Range: throughout Alaska.
Harvesting directions: Pick *surah* leaves in early spring, when bright green and sweet. Peel shoots of the felt-leaf willow, keeping the sweet inner bark and discarding the outer bark and woody core.
Food use: Nibble *surah* leaves as a snack, or add to salads. Leaves also blend well in herbal casseroles. The inner bark can be dried and ground as a flour substitute. Willow leaves are higher in vitamin C than oranges.
Medicinal use: Willow contains salicin, a natural aspirin substitute. If troubled by headache while hiking, you can chew willow leaves. (The rule of thumb is the more bitter the leaves, the higher in pain-relieving compounds.) You can also chip the bark, boil it in water, and sip the dark brew. For insect stings and bites, chew willow leaves and place the pulp on the irritated area. Use bark decoctions as a wash for wilderness wounds.
Other: Add willow to footbaths for sore feet. Use willow stems in basketry, and use leafy branches in the sauna to stimulate circulation. The pussy willows of spring make an attractive table decoration.

POISONOUS PLANTS

Following is a listing of toxic plants referred to in this book. Other poisonous plants also occur in Alaska. Before consuming any plant, be positive of identification. Consult medical help immediately if poisoning occurs.

ARROWGRASS
(*Triglochin* spp.)

Green leaves contain hydrocyanic acid, which can cause headache, heart palpitations, and even convulsions. Flowering stems of seaside species (*T. maritima*) can reach 2½ feet. Flowers are in dense spikes. Leaves are narrower and more rounded than goosetongue (*Plantago maritima*) and are most often mistakenly eaten in early spring.

BANEBERRY
(*Actaea rubra*)

The only deadly toxic berry in Alaska. A perennial, averaging 1½ to 3½ feet in height, with toothed, compound leaves and clusters of shiny red *or* white berries with a black dot at the end. Fruits are extremely bitter. All parts of the plant are toxic. Ingestion can cause sharp pains, bloody diarrhea, and even death due to cardiac arrest or respiratory paralysis.

DEATH CAMAS
(*Zygadenus elegans*)

Similar to a wild onion in appearance but highly poisonous! All parts contain zygadenine, an alkaloid that can cause salivation, muscular weakness, impaired breathing, and coma. Leaves are long and narrow; flowers are greenish-white with 3 petals and 3 sepals that look alike, giving the appearance of 6 "petals". Leaves and bulbs lack an onion odor. Death camas thrives in open woods and grassy places.

photo by Norma Wolf Dudiak

FALSE HELLEBORE
corn lily (*Veratrum* spp.)

This red lily contains a number of toxic alkaloids. It grows to human height and bears pleated broad cornlike leaves. The central, tall flowering stalk has many greenish flowers. Flowers have 3 petals and 3 similar sepals. Symptoms of poisoning range from numbness of the extremeties to diarrhea and stomach cramps.

photo by Edward Schofield

MONKSHOOD
wolfbane (*Aconitum delphinifolium*)

A rival of poison hemlock for toxicity. Contains aconite, which paralyzes the central nervous system. Flowers are purplish with darker veins (some specimens have white petals), and shaped like a monk's hood. Leaves are palmate, similar to those of wild geranium, but generally with narrower lobes.

POISON HEMLOCK WATER HEMLOCK
(*Cicuta* spp.)

These deadly plants contain cicutoxin, which depresses the respiratory system. Without prompt treatment, death generally occurs within 8 hours of ingestion. Hemlock bears white umbels, hollows stems, and chambered roots. Depending on the species, toothed leaves may be long and narrow or egg-shaped.

WILD CALLA
(*Calla palustris*)

Contains calcium oxalate, which can cause intense burning of the mouth and throat. Leaves are similar to those of marsh marigold, but come to an abrupt point. The plant bears a central flower spike (spadix) backed by a white modified leaf (spathe). Flowers are followed by clusters of red berries.

photo by Norma Wolf Dudiak

COOKING WITH WILD PLANTS

SALADS

Nettle Salad, Greek Style *Rosemary Gladstar Slick*

4 cups spring nettles
Marinade:
1 cup olive oil
½ cup lemon juice
Garlic, to taste
Honey, to taste
Toasted sesame oil, to taste
Garnish:
Feta cheese
Black olives
Sliced red onion

Steam nettles lightly, drain, and cool. (Cooking nettles takes the sting out.) In a blender, combine marinade ingredients and blend until creamy. Pour dressing liberally over greens and marinate two hours or more. Garnish with feta cheese, black olives, and slices of red onion. Serves 2 to 4.

Alaria Salad *Mrs. Wally Linn and Anne Wieland*

4 cups water
2 cups ribbon kelp (Alaria)
1 or 2 carrots, diced
1 or 2 green onions, sliced thinly
Grated ginger, to taste
Minced garlic, to taste
2 to 4 tablespoons rice vinegar
1 to 2 tablespoons soy sauce
1 to 2 tablespoons sesame oil
1 teaspoon sugar

Collect *Alaria* at low or minus tide. Gather only enough for one meal, as *Alaria* does not keep well. Pinch the blade just above the ribbon-like parts near the holdfast (these reproduce the algae). Discard the older, farthest out part of the blade that is tough or tattered. Rinse the younger, more tender remaining part of the blade.

Boil water. Add *Alaria* and blanch 1 minute. *Alaria* will turn emerald green. Drain off water. Cut *Alaria* into bite-size pieces. Add carrots, green onion, ginger, and garlic. Season with rice vinegar, soy sauce, sesame oil, and sugar. Toss salad and serve. Serves 2 to 4.

Floral Salad

Edible flowers of your choice: geranium, columbine, jewelweed, monkeyflower, violets, etc.

RECIPE CONTINUED NEXT PAGE

Edible in-season wild and garden greens
Grated beets and carrots for added color

This salad is highly flexible, adaptable to available greens and the size of your group. Neatly cover a platter with your greens. Then make circles of edible flowers and grated vegetables. If done with care and a bit of artistic flair, this yields a very attractive edible centerpiece. Try topped with thin *Wild Herb Pesto* (see recipe below).

Nine-Grain Salad *Darcy Nagle*

> 2 cups water
> 1 cup raw nine-grain mixture (brown rice, oats, red winter wheat,
>   rye, triticale, barley, buckwheat, sesame seeds, amaranth)
> ½ cup chopped lovage
> ¼ cup chopped wild chives
> ¼ cup chopped water chestnuts or ribbon kelp midribs
> ½ cup diced green and red pepper
> *Dressing:*
> ½ cup vegetable oil
> ½ cup tamari or soy sauce
> 3 tablespoons wine vinegar
> 2 teaspoons Dijon mustard
> 2 cloves garlic, pressed

Bring water to rolling boil. Add nine-grain mixture. Cover tightly and reduce heat. Steam for 25 minutes or until water is absorbed. Cool. Add vegetables and stir well. Blend dressing ingredients together and pour over salad. Refrigerate well. Serves 4 to 6.

DRESSINGS AND SAUCES

Wild Herb Pesto *inspired by Susun Weed*

> 1 cup olive oil
> ¼ cup pine nuts
> 3 cloves garlic
> 2 cups chickweed tips
> ¼ cup lamb's-quarter
> ¼ cup lovage
> 4 tablespoons sorrel
> ¼ cup Parmesan cheese
> Soy sauce, if desired
> Cayenne pepper, if desired

Blend oil, pine nuts, and garlic in blender until creamy. Add herbs and continue blending to make a thick green paste. Add Parmesan cheese and blend. Spike with a dash of soy or cayenne if desired. I vary this pesto throughout the green season, adding in-season wild herbs like shepherd's purse and goosetongue, and cultivars like basil and nasturtiums. Make a thinner consistency to use as a salad dressing, or a thicker paste for a cracker spread, pasta sauce, or dip.

Makes about 2 to 3 cups. For year-round enjoyment, pour thick pesto into patties, like pancake batter, onto a Saran-wrapped cookie sheet, cover, and freeze. When frozen, transfer to freezer bags for individual use.

Herbal Oil

> About 1 handful each *lovage leaves, mountain sorrel or sheep sorrel leaves, and chive leaves and flowers*
> *3 to 4 cloves garlic, chopped*
> *Olive oil*

Coarsely chop herbs and place in clean wide-mouth glass jar. Add garlic. Cover completely with olive oil. Let stand 2 to 3 weeks. Strain and rebottle oil. Use for salad dressings or as a cooking oil.

SNACKS AND APPETIZERS

Spicy Kelp Rings

> *2 cups peeled, sliced kelp stipe*
> *1 cup salsa*
> *1 tablespoon chopped chili peppers (optional)*

Marinate kelp rings in salsa and chilis for 24 to 48 hours, stirring occasionally. Remove from marinade and dry in vegetable dehydrator or oven with pilot light for 2 to 4 days. Store in glass jar. These spicy rings are popular as a snack or an appetizer. Serves 4 to 6.

Fiddleheads with Curry Dip

> *1 cup fiddleheads, cleaned*
> *¼ cup mayonnaise*
> *1 teaspoon curry powder*

Steam fiddleheads 5 minutes. Drain and chill. In small bowl, mix mayonnaise and curry powder. Chill dip to blend flavors. Serves 2 to 4 as an appetizer.

Pre-Dinner Puffballs *Harriette Parker*

> *1 teaspoon garlic, chopped*
> *1 tablespoon Parmesan cheese*
> *1 teaspoon sugar or less, to taste*
> *1 cup flour*
> *1 egg*
> *1 cup whole milk*
> *1 pound puffballs, peeled and sliced thin*
> *Butter*
> *Oil*

In bowl, mix well garlic, Parmesan cheese, sugar (to taste), and flour. In separate bowl, lightly beat together egg and milk. Dip puffball

RECIPE CONTINUED NEXT PAGE

slices into egg/milk mixture and then into flour mixture to coat. Heat 2 parts butter to 1 part oil (as needed) in a skillet, add sliced puffballs, and sauté slowly until puffballs are golden brown. Remove from heat and serve promptly. Serves 4 to 6 as appetizers.

Wild Wontons

> ½ cup wild chives
> ½ cup peeled cow parsnip stems
> ½ cup chickweed
> 1 cup lamb's-quarter
> 1 cup goosetongue
> 4 tablespoons Herbal Oil (see recipe, page 83)
> 1 package wonton wrappers

Dice cow parsnip stems and chop other herbs coarsely; sauté in herbal oil for 3 to 5 minutes. Place a spoonful on each wonton sheet and fold. Fry wontons as an appetizer, steam with goosetongue as a special vegetable dish, or simmer in soups. Makes approximately 3 dozen wontons.

SOUPS

Spring Soup *inspired by Martha Ellen Andersen*

> 2 to 3 cups nettle tops
> ¼ cup young green, nonfertile field horsetail
> 1 large onion, chopped
> 1 clove garlic, chopped
> 2 medium potatoes, chopped
> 2 tablespoons olive oil
> 4 cups chicken or vegetable stock
> 4 tablespoons chopped chives
> 3 tablespoons chopped lovage or oysterleaf
> ⅛ cup chopped dandelion leaves
> ½ cup half-and-half or buttermilk
> Herb salt, or salt and pepper, to taste
> Chives, edible flowers, or croutons, for garnish

Chop nettles (wear gloves!) and horsetail, discarding any tough stems. In a large pot, sauté onions, garlic, and potatoes in olive oil. Add stock, chives, and lovage or oysterleaf and heat to boiling. Reduce heat and simmer 10 minutes. Add horsetail, nettles, and dandelion and cook an additional 5 minutes or until potatoes are tender. (Cooking nettles takes the sting out.) Transfer mixture to blender and blend at high speed to liquefy. Add half-and-half or buttermilk and blend. Return soup to pot to warm. Season to taste. Serve in bowls garnished with chives, edible flowers, or croutons. (Seafood lovers can add shrimp sautéed in garlic butter just before serving.) Serves 3 to 4.

Plantain Miso Onion Soup

   1 onion, chopped
   1 teaspoon olive oil or toasted sesame oil
   2 tablespoons miso paste
   3 cups water
   ¼ cup chopped young plantain leaves
   ½ cup chopped chickweed tips
   4 tablespoons chopped chives
   4 tablespoons dry dulse or ribbon kelp
   1 package ramen noodles
   Parmesan cheese
   Wild Wontons *(see recipe, page 84) (optional)*

In a cast-iron skillet, sauté onion in oil, stirring frequently to brown. Add miso paste and water and stir. Heat to boiling. Add greens, sea vegetables, and ramen noodles. If desired, add *Wild Wontons* and ramen seasoning packet. Simmer 5 minutes. Top soup with Parmesan cheese and serve. Serves 2 to 3.

Fish Soup *Vera Angasan*

   4 cups water
   3 or 4 potatoes, diced
   1 large onion, chopped
   2 carrots, chopped
   ½ cup cooked brown rice
   1 clove garlic, chopped
   1 handful each *coarsely chopped mare's tail, sweet gale, beach peas, and dock*
   1 small handful coarsely chopped lovage leaves
   2 fireweed leaves
   2 cups salmon fillets, cubed

In a large pot, combine all ingredients except salmon. Bring to boil and boil hard 15 to 20 minutes. Add salmon and simmer 15 minutes, or until salmon is cooked. Serves 4.

MAIN MEALS

Tofu Medley *Molly Lou Freeman and Eva Saulitis*

   1 large yellow onion, sliced
   Garlic (as many cloves as desired), chopped
   1 pound mushrooms, sliced, or whole if small
   2 or 3 carrots, sliced
   2 tablespoons olive oil
   Rosemary, to taste
   Black pepper, to taste
   2 pounds tofu, cubed
   1 teaspoon tamari or soy sauce, to taste

RECIPE CONTINUED NEXT PAGE

1 large handful beach greens
1 small handful fresh lovage leaves
1 or 2 handfuls fiddleheads
½ cup Cheddar cheese, grated
Nutritional yeast (½ cup, plus or minus, to taste)

Sauté onion, garlic, mushrooms, and carrots in olive oil until onions are soft and clear. Add rosemary and black pepper to taste. Add tofu cubes and enough tamari or soy sauce to lightly steam everything. When tofu is warm, add the fresh greens. Toss and continue cooking until greens are tender. Add grated Cheddar cheese. When cheese is melted, add nutritional yeast. Serve with rice or noodles. Serves 4.

Spanakopita *inspired by Rosemary Gladstar Slick*

1 package phyllo dough
¼ pound butter, melted
14-16 cups wild edible combination of your choice: nettles,
  chickweed tips, lamb's-quarter, goosetongue, oysterleaf, etc.
4 eggs, beaten
1 pint cottage cheese
1 8-ounce package feta cheese

Using pastry brush, butter bottom of 9-by-13-inch baking dish. Place six leaves of phyllo on the bottom, buttering each with pastry brush. Chop herbs coarsely. Place in bowl and blend with eggs and cheeses. Spread half of mixture on phyllo base. Top with six more phyllo leaves, buttering each leaf. Spread remaining herb mixture. Top with six more leaves, buttering each. Drizzle remaining melted butter on top. Bake at 375 degrees for 30 minutes. Serves 6 to 8.

BEVERAGES

Tutti Frutti Tea

½ cup pineapple weed flowers
½ cup strawberries, sliced
½ cup lingonberries or bog cranberries
¼ cup raspberries, mashed
¼ cup raspberry leaves
½ cup fresh pineapple, sliced (optional)

Dry ingredients, as available, in herb drier or in gas oven with pilot light for 24 to 72 hours. Store in airtight jar. Place 1 heaping teaspoon per serving in prewarmed teapot, add 1 cup boiling water per serving, steep 5 minutes, strain, and serve. Makes approximately 1 cup dry tea blend.

Mint Julep *Marilyn Kirkham*

1 tablespoon fresh mint leaves or 1 teaspoon dried
1 cup boiling water

1 cup white wine (nonalcoholic if desired)
1 cup sparkling mineral water or carbonated water

Pour boiling water over mint leaves and let steep 5 minutes. Strain and cool. Add chilled white wine and sparkling mineral water or carbonated water. Pour into wineglasses and garnish with sprig of mint. Serves 3 to 4.

V-6 Juice

3 cups tomato juice
2 tablespoons each coarsely chopped lamb's-quarter, sorrel, and chickweed
1 tablespoon each coarsely chopped dandelion and plantain

Blend ingredients well in electric blender. Spike juice with dash of Tabasco if desired. Serves 3.

Gagne's Grog *Peter Gagne*

2 cups currants and other available edible berries
4 cups water

Simmer berries in water for 30 minutes. Drain, discarding berries and reserving liquid. Serve hot, spiked with rum, honey, or cinnamon stick if desired. Serves 4.

Slothhopper Smoothies *Jane Bell and Steve Johnson*

½ cup milk, soy or regular
4 tablespoons yogurt
1 teaspoon (or more) spirulina powder
1 teaspoon ground cardamom
½ teaspoon freshly grated nutmeg
½ teaspoon ground cinnamon
2 cups fresh or frozen wild berries
3 cups frozen bananas

Place milk and yogurt in blender. Add spirulina, cardamom, nutmeg, and cinnamon and mix well. Add fruits and blend until smooth. Enjoy this energizing health smoothie with a spoon, or add extra milk for a "shake" consistency. Serves 4 to 6.

BREADS AND BUTTERS

Earth Bread

1 onion, chopped fine
2 tablespoons oil
¼ cup ground clover blossoms
¼ cup ground lamb's-quarter leaves and seed
2 tablespoons ground nettles
1 cup flour

RECIPE CONTINUED NEXT PAGE

1 cup warm water
1 teaspoon kelp powder or sea salt

Sauté onion in the oil until brown. In bowl, blend sautéed onion, herbs, flour, water, and seasoning. Mix well. Add extra flour or water as needed to make a pliable dough. Knead on lightly floured board. Pinch off quarter-cup pieces and roll out into thin circles. Brown on both sides on lightly greased griddle. Makes eight to ten 6-inch rounds.

Wild Herb Butter *Eva Saulitis and Molly Lou Freeman*

½ pound butter or margarine, softened
8 cloves fresh garlic (more or less according to taste), chopped
¼ to ⅓ cup chopped fresh lovage leaves
⅓ cup nutritional yeast

Mix garlic, lovage, and yeast into soft butter. Enjoy on steamed beach greens, goosetongue, nettles, or fiddleheads. This is also delicious on potatoes, bread, or cornbread.

DESSERTS

Lingonberry Egg Rolls *Vickie Alto*

12 egg-roll wraps
Oil for frying
Wild berry filling:
4 cups lingonberries
1 cup honey
½ cup bran or oats
½ cup flour

Combine filling ingredients in saucepan and cook over low heat, stirring frequently, until thick. Divide filling among egg-roll wraps. Fold wraps according to package directions and fry in hot oil until golden brown. (These can also be baked in an oven for healthier, though drier, fare.) Makes 12 rolls.

Wild Berry Date Cake *inspired by Pat Sprintall*

¾ cup wild berries of your choice
¾ cup chopped dates
¼ cup butter
½ cup honey
1 teaspoon baking soda
½ cup yogurt
1¼ cups flour
1 teaspoon baking powder

Heat together berries, dates, butter, and honey until butter is melted. Remove from heat, stir in baking soda and yogurt. Sift together flour and baking powder and stir into mixture. Mix well and spread in greased 5-by-9-inch bread pan. Bake at 350 degrees for 35 minutes or until firm on top.

Floral Frosting

> 1 8-ounce package cream cheese, softened
> ½ cup honey
> ⅓ cup butter, softened
> 1 teaspoon maple syrup
> ½ cup edible flowers: fireweed, violets, roses, monkeyflowers, etc.
> 1½ cups milk powder

Cream together cheese, honey, and butter. Stir in maple syrup and 2 or 3 tablespoons of the flowers. Add milk powder, stirring well until mixture is smooth. Spread on cake and garnish with remaining flowers.

Barb's Blueberry Cheesecake *Barbara Sax Seaman*

> ### Crust:
> 1 cup butter
> ⅔ cup sugar
> ½ teaspoon vanilla
> 2 cups flour

Cream butter and sugar. Add vanilla and flour and blend well. Press mixture into the bottom and three-quarters of the way up the sides of an 8-inch or 9-inch springform pan.

> ### Filling:
> 2 8-ounce packages cream cheese (1 cup sour cream may be substituted for one of the packages of cream cheese)
> ½ cup sugar
> 2 teaspoons vanilla
> 2 eggs

Beat ingredients until smooth. Pour into crust.

> ### Topping:
> 2 cups blueberries (or wild berries of your choice)
> ½ cup sugar
> ½ teaspoon cinnamon
> 3 heaping tablespoons tapioca
> 1 tablespoon water

Simmer berries, sugar, cinnamon, and tapioca in water for about 5 minutes. Let mixture cool and pour over cream cheese filling. Push crust down with spoon around edges of the cheesecake. Bake at 450 degrees for 10 minutes and then at 400 degrees for 25 to 40 minutes or until firm.

GLOSSARY

Alga (*plural,* **algae**): any of a number of chlorophyll-containing organisms that live in fresh or salt water. Algae may be one-celled or many-celled, such as those plants we commonly call "seaweed."

Alkaloid: a nitrogen-containing organic compound that is insoluble in water. Alkaloids are generally bitter and usually potentially toxic.

Catkin: a male or female, drooping, spikelike cluster of flowers.

Decoction: an herbal preparation made by boiling an herb in water. Decoctions are often used with bark, roots, large seeds, and other hard plant materials.

Demulcent: a substance that soothes mucous membranes and other irritated tissues.

Dermatitis: inflammation of the skin.

Floral essence: a floral infusion used as a vibrational catalyst for emotional or spiritual healing.

Habitat: the type of environment in which organisms, such as plants, occur.

Homeopathy: a branch of medicine based on the "law of similars" in which minute quantities of substances (specially prepared through shaking and dilution) are used to eliminate symptoms that the same substances (if given on a large scale) would otherwise cause in a healthy person.

Infusion: an herbal preparation made by pouring boiling water over an herb and steeping; often used with leaves, flowers, and small seeds.

Liniment: an external herbal preparation prepared by steeping herbs in a solvent such as rubbing alcohol or brandy.

Pinnate: a compound leaf with leaflets arranged on two sides of a stem or long axis.

Plaster: a pastelike herbal mixture applied to the body to promote healing.

Potherb: a cooked vegetable.

Poultice: an external herbal preparation made by crushing or bruising plants and applying them (sometimes heated and/or wrapped in cloth) directly to the skin.

Purgative: an agent that promotes bowel evacuation.

Species: a group of plants that interbreed freely and have many characteristics in common.

Tincture: an herbal preparation made by steeping an herb in a solvent such as brandy, glycerin, or vinegar.

Tonic: a substance, such as an herb, used to strengthen body systems (often of a preventive nature).

Umbel: an umbrella-shaped flower in which all the flowers arise from one point.

Vermifuge: an agent that expels intestinal worms.

READING LIST

ALASKA

Fortuine, Robert, M.D., M.P.H. *Alaska Medicine,* "The Use of Medicinal Plants by the Alaska Natives." Vol. 30, No. 6, November/December 1988.

Hultén, Eric. *Flora of Alaska and Neighboring Territories.* Stanford: Stanford University Press, 1968.

Jones, Anore. *Nauriat Nigiñaqtuat, Plants That We Eat.* Kotzebue, Alaska: Maniilaq Association, 1983.

Kari, Priscilla Russell. *Tanaina Plantlore, Dena'ina K'et'una.* Alaska Region: National Park Service, 1987.

Pratt, Verna E. *Field Guide to Alaskan Wildflowers.* Anchorage: Alaskakrafts Publishing, 1989.

*Schofield, Janice. *Discovering Wild Plants.* Seattle: Alaska Northwest Books, 1989.

Viereck, Eleanor G. *Alaska's Wilderness Medicines.* Seattle: Alaska Northwest Books, 1991.

EDIBLE AND MEDICINAL PLANTS

Brown, Tom, Jr. *Tom Brown's Guide to Wild Edible and Medicinal Plants.* New York: Berkley Books, 1985.

Hoffman, David. *The New Holistic Herbal.* Rockport, Mass.: Element Inc., 1990.

Keville, Kathi. *The Illustrated Herb Encyclopedia.* New York: Mallard Press, 1991.

Moore, Michael. *Medicinal Plants of the Mountain West.* Santa Fe: Museum of New Mexico Press, 1979.

———. *Medicinal Plants of the Desert and Canyon West.* Santa Fe: Museum of New Mexico Press, 1989.

Rose, Jeanne. *Herbs & Things.* New York: Perigee Books, 1972.

* See *Discovering Wild Plants* for a comprehensive bibiography.

ACKNOWLEDGMENTS

This book is made possible by the countless centuries of earth's people, whose collective wisdom lives on with us in oral and written tradition. Special thanks to my friends, students, and fellow herbalists, who contributed such wonderful insights and recipes; to Ed Schofield, whose loving nurtures all aspects of my life; to photographer Norma Wolf Dudiak, for her persistence and excellence despite ever-changing lists and uncooperative weather; to Ed Berg and Jane Middleton for botanical and algological critiques; to Suellen Christiansen, for uplifting affirmations during deadlines and doldrums; and thanks to Alaska Northwest staff members Marlene Blessing, Carolyn Smith, and Cameron Mason.

HERBAL DIRECTORY

HERBAL CLASSES, SCHOOLS, APPRENTICE PROGRAMS
California School of Herbal Studies, Box 39, Forestville, CA 95436; programs range from one day to one year

Gardensong Herbs, Janice Schofield, P.O. Box 15213, Fritz Creek, AK 99603; day and weekend herbal classes, apprentice program

Good Earth Garden School, Ellen Vande Visse, HC02 Box 7471-C, Palmer, AK 99645; classes in organic gardening and communicating with nature

Sage, Cathy, 5108 Strawberry Rd, Anchorage, AK 99502; classes in culinary herbs

SAGE, Rosemary Gladstar Slick, P.O. Box 420, East Barre, VT 05649; classes and herbal correspondence course by founder of California School of Herbal Studies and Traditional Medicinal Teas

HERBAL PRODUCTS, FLOWER ESSENCES
Alaska Flower Essence Project, P.O. Box 1369, Homer, AK 99603-1369; flower essences and consultations

HerbAlaska, Marina Schaum, P.O. Box 15223, Homer, AK 99603; Alaskan lip balm, seasoning, teas, oils

HerbPharm, P.O. Box 116, Williams, OR 97544; herbal tinctures

Wild Weeds, P.O. Box 88, Redway, CA 95569; herbs, herbal products, containers, beeswax, essential oils

HERBAL ASSOCIATIONS AND PUBLICATIONS
American Herb Association, P.O. Box 1673, Nevada City, CA 95959; quarterly newsletter

Alaska Native Plant Society, P.O. Box 141613, Anchorage, AK 99514; native plant walks and newsletter

Alaska Mycological Society, P.O. 2526, Homer, AK 99603; mushroom walks and newsletter

The Business of Herbs, RR2, Box 246, Shevlin, MN 56676-9535; resource for small business owners

The Herb Companion, 201 E. Fourth Street, Loveland, CO 80537; articles on herbal gardening, culinary herbs, and herbal crafts

Wild Foods Forum, 4 Carlisle Way NE, Atlanta, GA 30308; newsletter on wild foods, and listings of events around the country

SEEDS
Abundant Life Seed Foundation (nonprofit), P.O. Box 772, Port Townsend, WA 98368; herbs, Pacific Northwest native plants, heirloom seeds, herb books

Baldwin Seed Co., P.O. Box 3127, Kenai, AK 99611; native wildflower seeds

INDEX

Boldface numbers refer to color photographs.

Alaska Northwest Books™ is proud to publish another book in its
Alaska Pocket Guide series, designed with the curious traveler in
mind. Ask for more books in this series at your favorite
bookstore, or contact Alaska Northwest Books™.

Alaska Northwest Books™
An imprint of Graphic Arts Center Publishing Company
P.O. Box 10306, Portland, OR 97210
1-800-452-3032